Suzie Edge trained as a molecular biologist before moving to clinical medicine, to spend more time talking to people, rather than just bugs in test tubes. She went on to work as a junior doctor in a variety of medical specialties including infectious diseases, haematology, and trauma and orthopaedic surgery. She completed an MLitt in Modern History to feed her fascination for the history of the human body and the history of medicine.

Always on the lookout for gory historical details, Suzie loves telling stories of how we have treated our human bodies in life and in death. She has over 470,000 followers on TikTok who love tuning into her stories of how famous monarchs met their end and human body stories throughout history.

T0173681

By Suzie Edge and available from Wildfire

MORTAL MONARCHS
VITAL ORGANS

vital organs

A History of the World's
Most Famous Body Parts

Dr SUZIE EDGE

WILDFIRE

First published in 2023 by
WILDFIRE
an imprint of HEADLINE PUBLISHING GROUP

First published in paperback in 2024 by
WILDFIRE
an imprint of HEADLINE PUBLISHING GROUP

3

Illustrations by Hayley Warnham

ISBN 978 1 0354 0461 2

Typeset in Adobe Garamond by CC Book Production

Printed and bound in Great Britain by Clays Ltd, Elcograf S.p.A.

Headline's policy is to use papers that are natural, renewable and
recyclable products and made from wood grown in well-managed forests and
other controlled sources. The logging and manufacturing processes are expected
to conform to the environmental regulations of the country of origin.

HEADLINE PUBLISHING GROUP
An Hachette UK Company
Carmelite House
50 Victoria Embankment
London EC4Y 0DZ

www.headline.co.uk
www.hachette.co.uk

For Derek

Contents

Introduction

Since the day I first picked up a scalpel under the bright lights of a medical school dissection room and cut into a human cadaver, I have been fascinated by the human body and all its parts. Everyone collects something and for years I have collected stories of body parts. (At least I'm not collecting actual body parts. Unlike like some people.)

From Napoleon's penis to Van Gogh's ear, from Marie Antoinette's teeth to Marie Skłodowska-Curie's bone marrow, this book brings together the remarkable stories of body parts that have made history.

In modern times, we like to think of the body as sacred, its rights inalterable, but despite what we like to think, we have always used and abused bodies. We've torn them apart, dug them up, experimented on them or taken bits home to display as trophies. Body parts have been used for propaganda in wars and have been pulled off in punishment. They've answered medical mysteries, been turned into relics and even saved lives.

As I started to think about bringing together these stories of organs, the news was full of Vladimir Putin and his invasion of Ukraine. News presenters and internet memes were questioning Putin's health, focusing in on his swollen face. This, they said, was a sure sign that the man was sick and that would explain his seemingly insane actions. It was obvious, they said, that he was undergoing medical treatment, judging by the look of him – chemotherapy probably, and likely use of steroids. There was no way of knowing whether these photographs, alarming as they were, were not photoshopped to make Putin look sick. To me, what was striking, apart from his swollen face, was that we were still looking to his body, to his health, to try to explain away or justify the atrocities we were otherwise struggling to understand. This was a scenario that has played out again and again throughout history; for as long as we have been telling human stories, we have used body parts to excuse the inexcusable.

As we will see, body parts have been instrumental in causing wars long before Putin. In 1739 the British fought the War of Jenkins' Ear (see page 93), where said body part was used to justify war with the Spanish in the West Indies. Even the First World War has been attributed by some in part to Kaiser Wilhelm's bitter reaction to his weak, withered left arm.

We claim to revere human remains. We feel that they should be treated with respect, as we want our own remains to be treated. Yet throughout history and often on the slightest whim we dig up, clean off and display human skulls as if they were merely interesting-looking stones we found on the beach. We keep mummified body parts stolen from colonised nations on dusty shelves in the back rooms of museums. We show off mummified specimens of the saintly in churches all over the world. Some body parts have been kept as trophies, prizes after a fight, or simply as a memento passed around the dinner table to fascinated guests. After all, who wouldn't want Napoleon's penis in a display cabinet?

In the eighteenth and nineteenth centuries, the removal of body parts, dried bones and pickled innards from the dead without anyone's consent, was a perfectly normal thing to do. Medical men on the one hand were keen on the study of the human body for future beneficence of their ailing patients, and on the other hand they gave very little thought to the idea that remains were anything other than specimens, meat and bone to be taken, shared, dissected and displayed. And it was not just medical men who would ship body parts about. Body parts were sent from all over the world when the wealthy visited foreign lands and treated the remains of the native people they encountered much as they treated native treasures and antiques. They put them in cabinets on display, much as they did with the heads of prey onto the walls of their grand estate houses. We can't judge every historical story on our twenty-first-century morals, but when it comes to stealing and displaying human remains I'm not convinced those morals have changed all that much.

My last book, *Mortal Monarchs*, charted 1,000 years of the English and Scottish crown by examining the deaths of their kings and queens. With so many fantastic stories about the exploits of royals, it was a challenge not to tell stories of the monarchs and their body parts here too, from William the Conqueror's exploding guts to Richard III's bones being put through the CT scanner. For now though, we leave those kings and queens alone and head off around the world in search of body parts that made history.

One monarch who did not fall within the scope of my first book and so luckily can be included here, is the French King Louis XIV. With all his well-documented illnesses and complaints, I could have written a whole book based on the vital organs and peripheral body parts of the Sun King alone. You can expect him to pop up often.

From head to toe, human beings just can't get enough of taking human body parts. We thankfully no longer tend to dig them up for display but the practice continues for other reasons. Bone dealers, blood farmers and organ traders take and sell body parts for financial gain. A worrying underworld of body parts trading, known as the Red Market, still operates today. In the past though, none of it was illegal.

Stories of body parts change how we see the world. All too often, historical figures feel distant and abstract, more myth and legend than real flesh and blood. The stories of their bodies and their parts, however, bring them back into focus and remind us that they were real, breathing creatures, who inhabited organs and limbs just like we do – until they were cut off, that is. Examining their bodies in this way can help us make a palpable connection to an otherwise difficult-to-imagine character. When we have a literal visceral response to a story, whether revulsion or reverence, it means that story and person have connected with us in some way. In the process of writing about their body parts, I feel I've come to know these historical figures in a different, more personal way than is usually offered by biographies, in keeping with my life-long fascination with the human body. I hope that by the end of the book you'll feel the same.

Emily Wilding Davison's skull

The Epsom Derby horse race, held on 4 June in 1913, was a huge event in England's society calendar. King George V and his wife Queen Mary attended and it was *the* place to be seen. Social classes mixed together. The King's horse, Anmer, and in particular its rider Herbert Jones, were easy to spot as they wore the King's distinctive colours of purple, red and gold.

As the horses raced round the track at speeds of up to 35 miles per hour, Emily Wilding Davison, who had been standing at the last corner before the final straight, ducked under the railings, moved away from the crowd and stepped into the path of the oncoming runners and riders. She reached out for Anmer, who made as if to jump the obstacle. It hit her at such speed and ferocity that she was bowled over. Her hat went flying one way and she rolled another. She came to a stop a few metres away. It was all over in a few seconds.

The news coverage of the event played in picture houses throughout the world. Viewers saw the grainy black-and-white footage of Emily stepping out and being hit by the horse. They might have spotted that she was holding something, but it was unclear. Had she successfully attached the scarf to the King's horse, it would have ridden off round the track trailing the suffragette colours of purple, green and white and displaying the words, 'VOTES FOR WOMEN'. Instead the scarf fell to the ground, as did Emily, her skull fractured.

Emily Wilding Davison was a suffragette, a vocal and militant member of the WSPU, the Women's Social and Political Union. This group of women sought equality, starting with gaining votes for women. At the beginning of the twentieth century, women had no vote in Britain. A vote would mean a say in political matters and decisions made in parliament. Women were just as affected by parliamentary decision-making, but they had no say in it and few political opportunities.

Davison was born in London in 1872 to an affluent family. She attended school and was able to go on to the Royal Holloway College who were promoting education for women in a time when such opportunities were scarce. But when her father died a couple of years later, she was left with nothing and it seemed she would have to abandon her education and work as a governess. Davison was determined enough to finish her studies by studying at home in the evenings. She saved enough of her earnings to attend St Hugh's College, Oxford, for long enough to qualify for the final exams. She earned a first-class degree but as women were not allowed to graduate and receive the degrees, she couldn't really finish what she had started.

Davison was frustrated by this inequality and wanted to do something about it, but there'd been little progress over the last thirty years or so. It's no surprise that Emmeline Pankhurst caught Davison's attention. Emmeline was an outspoken suffragette, whose motto was 'Deeds not words'. Emily Davison joined the suffragette movement. She was creative and imaginative and came up with new ways of causing disruption, including the burning of postboxes. She was also rumoured to have set fire to a house that Lloyd George was building. She threw rocks through windows and was found in the air ducts in the Houses of Parliament listening to proceedings. She spent the night of the census in 1911 in a cupboard in the Palace of Westminster, being discovered the next day by a cleaner. She was trying to escape the census count; why should she be included if she was not allowed to take part in elections? She was written on the census as having been in Westminster that night and in 1990 a commemorative plaque was placed in her cupboard.

Emily and other suffragettes were jailed for their militant

actions, and when they went on hunger strike in protest, they faced horrific interventions by their jailers to keep them alive. They were force-fed, which was brutal, akin to torture. Prison wardens, officers and even doctors used an instrument to prise the women's mouths open. They then forced a tube down the throat, risking it going into the trachea where fluids and food might be accidentally shoved into the lungs, risking their lives. If the women fought too much, a feeding tube was rammed down their nostril. This haunting torture was carried out twice a day for days on end. Emily fought on. In 1912 she was jailed for once again 'setting fire to postboxes, disrupting and demanding attention'. During one imprisonment she was not on hunger strike but was force-fed anyway. That was simply torture. This time Emily understood that more action was needed. Deeds not words. She threw herself off a balcony in the prison. It was a drop that might have killed her, but she lived, breaking her back and injuring her head. She wrote that had she died in prison people would surely have to sit up and notice the 'horrible torture that women faced'. But perhaps it would all just be covered up. She needed to do more. How could she be spectacular, in a way the government couldn't deny?

The more the suffragettes fought, the more the establishment fought back. Whilst they had male allies, there was also an anti-suffragette movement of men and women who would hold meetings and debates to defend the status quo. Some believed Emily's violence would only set back the campaign by alienating the public, rather than accelerate it. In the meantime, Emily wrote about what needed to be done to further the cause. She spoke of sacrifice and martyrdom. She was building up to something big.

On the day of the Derby, Davison saw an opportunity for a public display that would hit the headlines. So there she was, in the crowd as the horses were coming; now was her chance. She stepped out to her death.

The story that we have been told is that Emily 'threw herself under the King's horse'. Words matter and these ones, it is widely agreed by modern historians, do not really describe what happened. This phrasing suggests Emily Davison intentionally sacrificed herself and risked the lives of Anmer and Herbert Jones. In reality, she left nothing to tell us her objectives. For all we know, she meant the scarf attached to the horse as demonstration enough and to hopefully walk away unharmed.

Regardless of her intentions, the facts are that Emily was struck by the horse and a blunt force trauma fractured her skull. A skull fracture at its base can be diagnosed by the characteristic bruising patterns that develop around the orbits that are known as 'panda eyes' or 'raccoon eyes'. These would have come out over the days she lay in the hospital bed after the collision and are highly suggestive of a fracture at the base of the skull. Bruising behind the ears has the wonderful name of mastoid ecchymosis but it is a worrying sign. It is also called Battle's sign, not because someone has been in the wars but because it was first described by the English surgeon, William Battle. When the skull is fractured, blood can sometimes be seen behind the ear drum. Clear fluid trickling from the ear is another indication. The clear fluid is not good news; it is cerebrospinal fluid that surrounds the brain and spinal cord. Cerebrospinal fluid dripping from anyone's ear means that blood vessels and structures behind the skull bones have been disrupted by the break. Base of skull fractures are also associated with cervical spine fractures and cranial nerve injuries (the nerves that come from the brain directly to the face and head rather than the spinal cord nerves).

It is most often the skull's temporal bone that is broken when a skull fractures, but it can be the sphenoid, occipital and frontal bones that all come together at the base of the skull. All of this can result from a blunt force trauma like, for instance, being struck by the hooves of a horse or hitting the ground hard.

Photographs of Emily Davison just after the incident show her lying on the ground, unconscious. The top of her head had been covered up but blood is running down her face. A doctor in the crowd, Dr Vale-Jones, was hoping for a day off to enjoy the races but he made his way over to offer medical assistance. He wrote an account of what happened next.

I found her suffering from concussion of the brain and heart failure and her life! ebbing fast ... I called for brandy or whisky, [it was unclear if this was for him or the patient] and a "policeman" brought me the latter [that was fast], but this had but little effect to save life. A nurse then came up to my assistance. I said to her she is slipping out fast, I must have hot water (none to be had) I then sent a "policeman" to obtain a "Thermos Flask" that contained very hot tea, which I could hardly stand the touch.

I then took the nurse's handkerchief & poured some of the contents on & applied it to the left wrist. The 2nd application had the desired effect of restoring the heart action.

It is not exactly the ABC structure of first aid we rush to give today, but to be fair to Dr Vale-Jones, Emily's injuries were severe. She not only suffered a skull fracture but also other internal injuries. A big racehorse travelling at 35 miles an hour can do a lot of damage to a human body.

Three news cameras caught the event on film. Recently the equestrian and television presenter Clare Balding investigated the footage with modern technology. She saw a different story to the one we had been told. She concluded that it would have been no accident that the King's horse hit Emily. She *did* target this horse. It looks like Emily is presenting the horse with something, trying to throw something on it rather than simply looking to be run over by it. Even if she had not meant to 'throw herself under the horse', she would have known she risked severe injuries and maybe death. This was going to be a big statement, right in front of the news cameras.

Whatever Emily's ultimate intentions, the damage inflicted on her skull was enough to affect the functioning of the brain underneath. She died a few days later in the hospital. She never regained consciousness and became the first martyr to the cause of women's suffrage.

The King and Queen both wrote about the event in their diaries. Interestingly it was Queen Mary who was less forgiving. 'She ruined the day,' the Queen complained. The main concern of the newspapers was for the welfare of the horse and the jockey, though neither had been severely injured. Even so, and despite that Herbert Jones the jockey was powerless to stop Emily, he never got over it. The inquest held on 10 June recorded that this was death by misadventure rather than suicide. Five thousand supporters accompanied the coffin as her body was taken home to Northumberland. Deeds not words.

The following year, war was declared and everything changed. Just five years later the Representation of the People Act of 1918 increased the male electorate in Britain and allowed, for the first time in modern-era national elections, select women to vote. The war had been a major influence on the decision but the threat of further militant action was also a factor. For Emily Wilding Davison it would only have been a start, for there were further inequalities to fight. There still are.

William Shakespeare's
and Lord Darnley's skull

Whilst Emily Wilding Davison's fractured skull is remembered here as a symbol of her sacrifice for her cause, other skulls have become memorable for arguably far less honourable intentions. Skulls are a provocative symbol of death. Though we don't see what skulls look like in life, the instantly recognisable face structure stares out at us from gravestones and signet rings as memento mori – a reminder that we too shall die.

When it comes to such a loaded symbol as a skull, it is hard to decipher true stories from gothic-horror fiction, even when it comes to the most notable of skulls. Every haunted mansion house in the land has a screaming skull story, a skull that will shout and wail if it is ever removed from the building in whose walls it is entombed. Other skulls have indeed been involved in post-mortem adventures, stolen as trophies or sold as curiosities. Though we like to think that our bodies shall rest in peace after our deaths, our skulls in particular have a habit of going missing and becoming fodder for legend-tellers. They are a symbol of what lies ahead, but not necessarily of what lies beneath.

There are many stories of skulls that do not rest in peace along with the rest of their remains. Instead, they rest in pieces, as it were. One such case was Henry Stuart, Lord Darnley, the second husband of the iconic Mary, Queen of Scots, and the father of James VI and I. He was a scoundrel and a murderer, and he met his end in rather strange circumstances. Mary and Darnley were staying at Kirk o' Field in Edinburgh, Scotland, in February 1567. She was out one night attending the wedding of a friend and servant. He was sick and stayed at the lodging. While she was partying, dastardly deeds occurred. Explosions were heard at Kirk o' Field, and when people went to investigate, they found Darnley and his valet, not having been blown apart but having been smothered to death. Their dead bodies lay in the garden grounds. Two barrels of gunpowder were exploded in the rooms below his chamber, but he had already fled the building and had escaped death by explosion. Evidently, the murderers caught up with him outside.

As to the identity of the murderers, there was little evidence to go on. Strewn around the bodies was a chair, a cloak and a dagger; it was quite literally all very cloak and dagger. Whoever was behind it, Darnley was dead and his body, as they do, was already starting to decay. He was embalmed, encased in lead and interred in the Royal Vault at the Abbey Church at the Palace of Holyroodhouse. The bodies in the vault survived various attacks over the years, but in 1768 the roof collapsed, and later the vault was exposed. Inside, were six lead coffins. Two were thought to hold the bodies of James V's infant boys, Mary's brothers. There was James V himself, blackened by the tar-like balsam used to embalm him, and there was his first queen, Magdalen, who was rather well-preserved. The Countess of Argyll, an illegitimate daughter of James V, was also interred there.

While the coffins were exposed, someone came along with a shovel and a bag. Darnley's skull was stolen, as was Queen Magdalen's head. Darnley's skull would come to be listed amongst other curiosities by the Antiquarian Society of Scotland. There was no record of how it had got there, suggests William Wright, writing in *Nature* in 1929, because the head was 'obviously nefariously acquired'. What happened next was that two heads turned up, both claiming to be Darnley's. Darnley might have been known to be two-faced, but this was taking it too far. One, or both of them, were fakes. Human skulls, yes, but was either of them actually Lord Darnley's? Who was to know? One of the skulls was held by the Royal College of Surgeons of England, in London. It had been presented to the RCS after being bought at a Sotheby's auction as part of the private collection of the Hon. Archibald Fraser of Lovat, whose own father had been beheaded for his part in the Jacobite uprisings. This was the skull that had been listed by the Antiquarian Society of Scotland. It was destroyed during the Blitz but images remained. The other skull, of origin unknown, was still in Scotland, held in Edinburgh at the University.

Before it was destroyed in the bombings of the Second World War, the RCS skull had been examined extensively. Skulls and bones can silently give so much information, even if it might be interpreted differently by different examiners. One notable observation was that the lack of any soil or earth within the cavities, such as might be seen in a skull that had been at some point buried. The skull of a hanged criminal who had bypassed the grave and gone straight from the gallows to the anatomists would look like this, without evidence of an earthly burial. Had this specimen even been in the grave? Another interesting aspect of the skull was some pitting that was evident on the surface. This, it was said, was a sign of his 'incontinence'. They were euphemistically suggesting syphilis. To be visible on the skill, such an infection would have to be developed enough to reach the tertiary stage. Another close examination disagreed, the pitting did not look like an inflammatory process; rather this was a deliberate act, using an instrument post-mortem, done to make the skull more interesting, more noteworthy. Another investigator thought perhaps that the pitting was the result of burrowing insects, hopefully also post-mortem. Sadly though, only images remain of that skull.

In 2016 Emma Price, a student learning the skills of facial reconstruction at the renowned Centre for Anatomy and Human Identification at the University of Dundee in Scotland, decided to find out. Computer reconstructions of what the face belonging to that skull would have looked like were compared to portraits. Price used the images of the destroyed London skull and examined the Edinburgh skull labelled to be that of Darnley. She argued that the RCS skull destroyed in London was more likely to have belonged to Darnley. The other was way off, with arched eyebrows and a distinctively sloping forehead that bore no resemblance to the portraits of the infamous rogue. Who that skull might have belonged to and how it ended up labelled as Darnley, nobody knows. Perhaps that's what became of Queen Magdalen's head, stolen at the same time as Darnley's?

It would appear we just like digging up skulls. For trophies? For ransom? For good luck? We have archaeological evidence that for centuries, throughout Europe, skulls have been turned into drinking cups. Often it was done in celebration of killing an enemy. Sometimes, it was just on a whim. When the remains of a monk were unearthed in the grounds of Lord Byron's house, Newstead Abbey, the poet did not respectfully rebury the cleric. Instead he made the skull into a goblet to drink from.

One skull that has sat apart from any related bones in a church crypt for hundreds of years has long been claimed to be that of none other than William Shakespeare. Sitting fifteen or so miles from where he was buried, this skull could have belonged to the man who dreamed up the plays and sonnets that are so much a part of our lives even four hundred years later. Surely, of all skulls, his would be most interesting to the phrenologists, the doctors who believed that pseudo-science of lumps and bumps on the head correlate to personality types and intelligence. Skulls became sought-after in the eighteenth and nineteenth centuries because of this practice, particularly those of the well-known and the talented, musicians and writers in particular. The skull of writer and political pamphleteer Jonathan Swift was stolen to order. Musicians Haydn and Beethoven's skulls were also taken from their graves. This century, the skull thought to belong to Shakespeare has been scanned and analysed and found to belong not to the Bard but to a woman in her seventies. It didn't really fit the criteria as Shakespeare died as a fifty-two-year-old man. So, that myth of it being Shakespeare's skull was laid to rest, even if the mystery of the unknown woman was just beginning.

There's something almost sad about how the use of sophisticated modern technology is changing our perceptions of these relics. Having proudly reported to be one thing for generations, they turn out to be something else entirely. What technology cannot tell us is how on earth the skull of an elderly woman came to be in a church for so long, just sitting there, claimed to belong to one of England's most famous sons.

When Shakespeare shuffled off this mortal coil, he was buried at the Holy Trinity Church in Stratford-upon-Avon where he was born and died. His grave was marked by a stone that warned anyone who harboured thoughts of interfering with it: *Cursed be he that moves my bones*. Shakespeare's concern was that his remains would end up in the charnel house, a vault where bones were stored after graveyards were cleared to make way for the newly deceased. It seems even being buried within the church walls may not have been enough to preserve his remains. Alas, poor William.

In 1871 an anonymously published story appeared in *The Argosy* magazine. 'How Shakespeare's Skull was Stolen by a Warwickshire Man' was a seemingly fictional portrayal of a young medical student named Frank Chambers, who had heard that a wealthy collector might pay good money for Shakespeare's skull, and as he had resurrectionist connections owing to his profession, he was able to get a gang together to dig up the head. When they did so, he could find no buyer and decided it should be put back. That's when the tombstone, which lies flat across the grave, was broken. This story was just that, a story, but when investigators with that pesky modern technology went to investigate, they found some remarkable similarities with the tale. A writer of fiction could not have known that Shakespeare was buried, not in a coffin in a vault, but in a shroud only three feet below the stones, buried in earth, and yet what they wrote corresponded with the findings of the investigators. At the head end of Shakespeare's grave, where archaeologist Kevin Colls and geophysicist Erica Utsi studied the area using ground-penetrating radar, there is no obvious head, but instead it looks like a repair job had been carried out right underneath the stone. Damage to the wall structures where the head should be had been repaired. The ledger stone that marks the grave is shortened. Perhaps it was broken during a raid. Shakespeare's head may or may not lie with the rest of him. If someone did take it, a curse could be upon them. The Church have stated that they have no intention of going digging and finding out if either the story of body-snatching or Shakespeare's curse are fact or fiction.

Ground-penetrating radar and modern facial reconstruction techniques like those used here are starting to pour cold water on these centuries-old myths and legends, but they are also digging up as many questions as they are burying. One modern analysis has shown that a legend could have been a true story all along.

William Longspee's twelfth-century skull is remembered for rather sinister reasons. Longspee (a nickname referring to his long sword) was the illegitimate son of England's King Henry II and his mistress, Ida de Tosny. That made him the half-brother of Kings Richard the Lionheart, and John. It was a risky time for anyone to be related to royalty in case they be considered a threat. Their young nephew Arthur did not even make it to adulthood; he was killed by a conniving John, who wanted the crown for himself.

For years, William Longspee fought alongside his half-brother Richard in France, but when The Lionheart was killed by the gangrenous wound of a crossbow bolt in 1199, Longspee returned to the cooler climate of England. There, he sided with his other half-brother John, but John was notoriously bad and he had enemies. When other barons turned against John, Longspee remained loyal. He remained loyal, too, to John's young son, Henry III.

In 1226, William Longspee died unexpectedly. He was nearly sixty years old but it was not immediately obvious what had killed him. Something other than natural causes must have struck down this strong healthy man, it was thought. Blame for his death was aimed at the influential and powerful Hubert de Burgh, the Earl of Kent. It sounded suspiciously like poison was involved. What could be done in thirteenth-century England with a dead noble and a conspiracy theory? Not a lot, and so Longspee was put in a box. He was the first person to be interred at the stunning Salisbury Cathedral. He had patronised this place of worship and laid its foundation stone six years before.

In 1791, Longspee's coffin was opened up for a look and the mummified remains of a rat were found inside his skull. It looks as if the rat came along, drawn perhaps to the yummy aromas of rotting flesh, and as the man started to decay, the rat went to lunch. Ratty munched his way through the soft tissue into Longspee's brain – perhaps through the foramen ovale, the large hole at the bottom where the nerves of the spinal cord come out – and carried on munching. If Longspee was poisoned then Ratty was eating into the poison within the remains.

The rat was well preserved. When it was analysed years later it was found to have died in the same way as his lunch, by eating arsenic. That's how Ratty's remains were in such good condition. Many centuries later, arsenic was used as an embalming chemical, but not in the thirteenth century when Longspee was laying foundation stones and fighting off the

French. Perhaps William Longspee was murdered after all. The rat is still well preserved in the museum in Salisbury. You can go and see him there, along with the tomb of Longspee, but whatever you do, don't lick the rat.

Jeremy Bentham's head

Whilst skulls have been stolen, dug up, bought and sold, exhibited as trophies and used as drinking cups, there are also whole preserved heads that have found their own place in history. That is, heads where the connective tissue, the muscles, and the skin are still attached to the bones, though it's not always in a pleasantly preserved manner.

If you have good enough reason to, University College London will permit you to see the nearly two-hundred-year-old preserved head of Jeremy Bentham. Be prepared though, that thing isn't pretty.

Jeremy Bentham was a philosopher, social reformer and the very definition of the posh Georgian eccentric. He is remembered for his ideas around utilitarianism, saying that it is the greatest happiness of the greatest number that is the measure of right and wrong. The philosopher is remembered for using his head in more ways than one.

Bentham was interested in education and thought about how we can be of best use to each other. He felt it was a shame that after death, bodies were not being put to good use. Very few cadavers went to the doctors and medical students for their learning (well, not officially anyway; thousands went unofficially). Otherwise they were left to rot and return to the earth without providing any value to those still alive and kicking. So he made plans for how our dead bodies might be put to use, whilst also reducing the need for so much graveyard space. What was this idea, you ask? Big museums full of dead people. Weird, right? So the problem might have needed a bit more thought.

Bentham decided that the best way to convince society to adopt his great idea was by putting his own dead body on display. Familiarity with corpses would 'diminish the horrors of death', he hoped. He called this idea the auto-icon, and he wished it would be taken up by everyone. The bones should be displayed with clothes stuffed with straw to look lifelike. The body could stay, perhaps in the corner of the living room or the office, to provide a reassuring presence. His head, he requested, should be preserved 'in the style of the New Zealander' and placed on top of the display. He asked his friend Dr Southwood

Smith to carry out his wishes. The doctor did so, dissecting his friend's corpse at the Webb Street Anatomy School when Bentham died in 1832. Bentham's body, however, had rather the opposite of the intended effect. The preservation process did not quite go to plan and distorted his facial features. Rather than lessening the psychological impact of death and decay, his mummified head resembled a house of horrors exhibit.

Smith had a wax model made in Bentham's likeness instead, and for a long time the real head was simply placed in a box that sat at the auto-icon's feet. It became a target, and there are stories of it being stolen, kicked around and being wheeled into council meetings to make a deciding vote (always to side with the motion). Students from King's College stole the head in 1975 and they held it to ransom, asking that the price be paid to a charity as part of rag week. After that, Bentham's head was no longer on display. Instead it was locked up, out of temptation's way. Now it's harder to access and not quite what Bentham wanted, but as per his wishes, it can still be seen if you have a strong enough stomach. If you decide to look on Google images at this juncture be warned of nightmares.

The attempt to preserve Bentham's head 'in the manner of the New Zealanders' fell flat on his, well, face. The result was too horrifying to keep on public display. So what did Bentham have in mind? He must have seen the Māori heads that were being collected and displayed in Britain and elsewhere during the nineteenth century. The Māoris had perfected their preservation process over generations of reverence for their ancestors. Copying the process turned out to be a tough job. The biggest difference, though, between Bentham's intended display and that of the Māori mokomokai heads taken away from New Zealand, was one of consent. Bentham wanted his head to be exhibited. The owners of the Māori heads had little choice or say in whether or not they would leave their ancestral homes.

Mokomokai heads belong to the Māori's tattooed ancestors. They were removed from the dead and preserved following a particular custom. At first, they cut away the brain and the eyes – the squishy bits that would putrefy and smell the fastest. Then they packed any orifice with flax fibres, gums and waxes. The heads were then baked or steamed, and after that they were dried in the sun. They were then coated in shark oil and the insides packed with clay. The heads were kept and the dead remembered, revered, honoured. The procedure preserved not only the ancestor's likeness, but importantly the facial tattoos the chiefs had earned throughout their lives. The marks represent rank, status, family, deeds and achievements. A high-ranking chief would have a face full of tattoos. Women, too, of high status would have the tattoos, perhaps only on the chin and lips. The heads of killed enemies were also preserved in this way but they would not be treated with the same respect, rather they were mocked and showed off as trophies of war.

The severed heads caught the eye of colonising Europeans who saw not tradition to be respected, but collectible items to bring home as curiosities and trophies. The first head removed from its homeland was believed to be that of a fourteen-year-old lad from the 1770s. The boy's family were not so keen to trade with Joseph Banks, the British naturalist on board HMS *Endeavour*, but a musket helped them make their decision, and so they parted with the sacred head in exchange for not being shot and for some underpants.

The trading of the mokomokai soon spread and hundreds were taken to Europe, Britain and the US, with the result that preserved heads were in short supply. Someone had the bright idea of just making new ones. Slaves and prisoners were selected, tattooed and murdered so that their heads could be traded or sold. These imitation heads made their way, alongside the authentic Māori ones, to the fashionable trophy cabinets and dinner parties of the west.

Over the last few decades, the Māoris have campaigned to have the heads returned to their homelands and they are starting to have success. France made the headlines by returning some mokomokai in 2010, but it was not all plain sailing. At the start of this century, a new curator took charge at the French museum at Rouen and when he looked through the archives held within, he came across a distinctive mummified head. Its markings showed it to be a mokomokai, the dried head of a

Māori, a native of New Zealand. He thought it would be best to return the head to New Zealand rather than let it gather dust in the museum storeroom. It certainly was not going to be on display any more, as by then it was considered inappropriate. There was a problem though. The government opposed his honourable plans as the law in France at the time was that museum pieces could not be returned. They were 'part of the culture': they said such heads were no longer considered human remains, but had transcended into artefacts of French cultural significance. Anyone looking at the face of a mokomokai, even with false eyes staring back, would be hard pushed to say that these are no longer human remains. Be that as it may, the French government stopped the return of the heads to New Zealand. They feared that this would be a slippery slope: that skeletons would be found in closets, literally, and the shelves of the museums emptied.

Then there was the case of the Khoikhoi woman Sarah Baartman, known as the Hottentot Venus. The name, now considered derogatory, was used to label not just Baartman but other African women, too, who were taken from their homes and exhibited in shows across Europe in the early nineteenth century. Baartman had a condition known as steatopygia, a large build of fat, particularly on the buttocks. It was not uncommon amongst her people, but in Europe it was seen as something to parade and mock. She was taken advantage of and abused and when she died in 1815, that abuse carried on. Her brain, skeleton and sexual organs were preserved for all to keep looking at in a Paris museum until as recently as the 1970s. In 2002, a few years after Nelson Mandela had requested the return of her remains, the French agreed to send them home, despite the previous rulings they had made regarding the Māori heads. The law was questioned once again and this time, in 2010, France's National Assembly voted to return sixteen Māori mokomokais from museums across France. Now over three hundred mummified heads have been returned to New Zealand, but many more have still to find their way home to be buried respectfully.

It was not just in New Zealand that heads were a symbol of triumph in war. In 1838 King Badu Bonsu II, ruler of the Ahanta tribe in present-day Ghana, decapitated two Dutch emissaries and hung their heads as decorations upon his throne. It is not known what they said to upset him. In revenge the Dutch decapitated the king. Nobody knew what happened to his head until it turned up recently in a Dutch museum, pickled in a jar of formaldehyde. The head was returned to the Ahanta tribe and to Ghana. Displaced heads all over the world are beginning to find their way back home.

Some heads, though, remain on proud display. Religious relics, it seems, are a law unto themselves. St Oliver Plunkett's head gives off similar ghoulish vibes to that of Jeremy Bentham's, but this one is actually on display in Ireland. This religious relic is regularly visited by school children. It is enough to give the youngsters nightmares, but perhaps it's worth it as the severed head resides in an elaborate gold shrine at St Peter's Church in Drogheda, County Louth.

Plunkett was hanged and drawn and quartered for high treason in 1681. He was one of many victims of the fabricated Popish Plot. The plot, concocted by the Englishman Titus Oates, spread fear and hysteria against Catholics and many were put to death for their alleged part in it, even though it turned out to be a hoax. Plunkett, who was the Archbishop of Armagh and Primate of all Ireland, was one such victim. He was condemned for 'promoting the Roman faith' and now his mummified head serves as a reminder of religious persecution. His quartered body was buried where he was killed but his head was sneaked away by nuns and taken to Rome. In 1921 it made its way back to Ireland and now tells us the story of centuries of religious turmoil.

The hunt for secular relics continued too. The settlers who colonised Australia in the nineteenth century took vast numbers of displaced remains and sent them all over the world.

The bodies of aboriginals were routinely cut up into different parts and became sought-after trophies reminiscent of the deer hung on walls and in cabinets of hunters. They were displayed in homesteads in Australia but others were sent abroad, to Britain, the United States and to Europe. Many went to institutions where they were sent in the name of education. A decade ago, the Australian government believed that the remains of as many as 900 indigenous Australians were still far from home. Those who died in institutions were dissected by medical students and doctors for their education, so many came from the hospitals, the jails and the asylums. Others were exhumed from their newly dug graves for the same purpose. In Europe and America, the pseudo-science of phrenology was all the rage and skulls became hot commodities. Aboriginal skulls were easier than some others to come by. Many of them went to private collectors, others ended up in the Smithsonian, who returned their collection to Australia in 2010. One of the skulls known to belong to an aboriginal was sent by Dr William Ramsay Smith to his friend at Edinburgh medical school. All sorts of body parts were sent by Smith to the Scottish medical school, either pickled or dried. Notably the skull displayed the unmistakable signs of a gunshot to the head.

As in New Zealand, Australians are now campaigning for the return of displaced heads. An Aboriginal warrior by the name of Kanabygal was shot and beheaded in 1816. His head was taken as a trophy and now sits in an archive at The National Museum of Australia in Canberra, among the remains of 725 other Aboriginal people. Aboriginal people were thought of as part of humanity's story, as the 'missing link'. Arnhem Land people were likened to Neanderthals and so their bodies were studied extensively. They were even referred to as 'Australian's stone age men' and so their bodies were picked apart, prodded, poked and paraded. Kanabygal is unique amongst them; he is the only one to have a name and that we know how his skull came to be part of the archives. Most of the others remain nameless and their individual stories can never be told. As a group, though, their collective story is now being told by the museum.

The last full-blooded man of the Ngambri people of Limestone Plains, OnYong, died in 1850. He was killed by spear. Well, not a spear, but a man holding a spear. He was buried, but it was not long before settlers, and a visitor called Smithwick, dug up his remains and took his skull. Smithwick fashioned it into a sugar bowl, some say. Other reports say an inkwell. OnYong's descendants believe that the sugar bowl (or inkwell) skull remains in a private collection in Canberra, Australia's capital. 'The purported collector has never returned my calls,' said Paul Daley, an Australian news journalist, in 2014. It might be some time before all the world's displaced body parts make it home, to be respected as Bentham's is.

Albert Einstein's brain

One Monday morning in 1997, about the time that Albert Einstein's stolen brain was making a journey across the United States, I stood in an anatomy lab looking down at a row of dissected slices of human brain. Fixed in formaldehyde and sliced about a centimetre thick, they were being arranged on the table as we joined that day's anatomy class. 'Gather round,' the anatomy instructor said to our group of white-coated students, and we all moved closer to the specimens. I was not at all steady on my feet. It was not the sight of sliced human brains that had me swaying though, but rather, I had been watching a rugby match the day before, accompanied by drinking games. My liver had been put to good use, breaking down the alcohol but not before the chemicals affected my brain and particularly my cerebellum.

The anatomy instructor was astute. It must have been obvious that I had very recently been under the influence and so she picked me to help with a demonstration. She took me through a series of intricate hand movements that should show off fine motor control such as pinching, pointing, tracing a wriggly line. I failed all the tests. My fine motor control had been knocked off – temporarily, I hoped. It was a teaching moment that has stuck with me. Though I said I would never drink again, I didn't keep my word. All in the name of education, of course. We turned to the brains on the table, we identified each part and we talked about the various functions. My own brain might have been groggy, but I was fascinated.

To access the brain, you saw the top off a skull, carefully peel away the visible meninges and look at the jelly-like blob underneath. This examination will reveal few obvious clues, however, about how the brain works. It is the trickiest of all human body parts to look at and derive its function. We see in bone structure the function of strength, stability and muscle attachment. We look at the chambers of the heart and the vessels coming off it and gain an understanding that it beats and is a pump for blood. By observing the lungs and the sacs attached to the air tubes, we see that this is a place for the movement of air, even if we know nothing of what that air contains or does. Yet we cannot lay bare the brain and immediately understand from its shape and structure that this is the place of our personalities, our decision-making, our autonomic functions and deliberate movements, our memories, and even our dreams.

Brains are so individual and yet, cut them out of the skull, pickle them in formaldehyde and slice them thick as a steak, what we would see is that yours looks much the same as mine. It is remarkable they are so similar considering that through neuroplasticity, they are shaped individually on a cellular level by our own very personal experiences of life.

Right now, your own neurons are firing in that part of the brain where reading and written-word recognition are undertaken. Or if you are listening in audio, another part of your brain is working hard at listening and language recognition. Perhaps you're feeling revulsion at the idea of our sliced pickled brains on a dissection bench. Or maybe you're not wired like that – it could be that your brain is instead thinking, *Well, that's cool.*

In 1861, Frenchman Pierre Paul Broca found that there was a specific part of the brain that dealt with speech. He had two patients whose brains had been damaged in just that area and both had lost their ability to speak. This area, in the frontal lobe of the dominant hemisphere, usually on the left, became known imaginatively as Broca's area. It led to the discovery of more parts of the brain that deal with specific functions.

People's brains would have to get damaged though, before observations could be made.

In 1848, Phineas Gage was a respected foreman building the railroads in Vermont. He survived a remarkable injury, but after that he wasn't respected quite as much. Gage's tamping iron had sparked into some gunpowder and the resulting explosion sent the javelin-like iron rod up through his face, behind his left eye and out through the top of his head. He did not lose consciousness. The doctors at first questioned if the iron had indeed gone through his skull, owing to his lack of dropping down dead. The first doctor pushed back bits of the brain that he thought might look 'good for something'. A second doctor cleaned things up better by sticking his fingers in both ends of the head hole and removing broken bits of bone and brain.

It is truly remarkable that Gage survived, but what came next was even more interesting. The accident and his survival exposed something extraordinary. His personality completely changed. He became really horrible to be around, and his friends said he was no longer the Phineas Gage they knew. His injury helped us understand that the personality, the ability to self-regulate and self-inhibition come from the frontal lobe, an area of the brain that in the case of Phineas Gage, had been damaged. People with frontal lobe injuries can become disin-hibited, saying things they would not have dreamed of saying out loud before. Or they do outrageous things they would not have done previously. A similar scenario occurs with Pick's disease, where the frontal lobe is destroyed by the neurodegenerative disorder. Those around sufferers of Pick's disease, a type of frontotemporal dementia, can see their loved ones change into a completely different person. They suffer more perhaps, than the person with the disease itself. We understand this pathology in part because of Gage's case.

Victorian phrenologists might have been wrong that the shape of the head determines personality types and thoughts, but it was the beginning of something. Now we understand that it is in fact distinct areas within the brain that deal with different functions.

We have liked the idea, over the years, of picking apart and analysing the brains of those we deemed different. We want to pin their differences on the anatomical structures, just as we have with specific functions thanks to Phineas Gage and the patients of Pierre Paul Broca. What would the brains of the exceedingly clever, like Albert Einstein, or of criminals hanged at the gallows, or the mentally ill, or of a dictator like Benito Mussolini tell us?

Born in 1879, Albert Einstein became a theoretical physicist. He is often said to be the smartest thinker of all time. Hear the word 'genius' and it's likely that the moustachioed, wacky-haired image of Albert Einstein pops into your mind. He gave us the general theory of relativity, bringing together four papers that he wrote about his thought experiments. He gave us theories in quantum physics that we all pretend to understand. Not even a genius can live forever though, not yet, and on 18 April 1955 Einstein died in the hospital at Princeton, New Jersey. He died from a ruptured abdominal aortic aneurysm. He was seventy-six years old. Einstein had known that there were people interested in getting their hands on his brain after his death. He was clear that he did not want his brain taken, dissected or displayed. He wanted to be cremated 'so that people will not worship at my bones'. He knew people would want to do that. So, Einstein might have understood the universe, but he also understood human nature just as well. His wishes were ignored. Within a few hours of his death, his brain was removed.

The pathologist on call was Thomas Stoltz Harvey. He must have been delighted when he took the call. Albert Einstein just happened to be his hero and he got to perform the autopsy. After establishing the cause of death, Harvey betrayed his hero's wishes. He kept Einstein's brain. He removed the eyes too, handing them to Einstein's ophthalmologist. The brain went into a cookie jar where it was stored – well, most of it was – for forty years. He sliced into it 240 times, sectioned it, prodded and poked at it, and he took photographs of it. He studied every tiny aspect and he sent several pieces to labs and scientists all over the US and the world.

What was he expecting to see? Different shapes? Different sizes? A little signpost on the brain's cortex: 'this way to genius'? Was he expecting to unearth something that could be recreated? Could we just pinpoint and activate genius mode? It didn't make the doctor any smarter, but various interesting observations were made.

It was noticed, when looking through the microscope, that the concentration of glial cells was higher than average in Einstein's brain. Glial cells form part of the structure of the central nervous tissue. They support the neurons, provide oxygen and nutrients, and produce the myelin that surrounds the nerves cells. Einstein's increased rate of these cells presumably put in more work to support his grey matter. Researchers also observed that in Einstein's brain, the corpus collosum had much thicker connections. The corpus collosum is a structure that connects the left and right hemispheres of the brain, allowing for increased communication between areas that deal with the likes of abstract thinking, decision-making, visual processing. Perhaps Einstein was born that way, and with these increased connections he was able to be so much more creative than us average thinkers. Or was it that Einstein's thoughts and work created a thickened corpus collosum because of neuroplasticity shaping his brain? The lesser-mentioned reality is that most of Einstein's brain was no different from comparison specimens, those that were deemed normal or average.

In the late 1990s an American writer called Michael Paterniti got in touch with Dr Harvey, who was by then in his eighties. They talked about the history of Einstein's brain and the pathologist expressed that he would like to give it a new home. Paterniti offered to pick up the doctor (and the brain) in his car, and Harvey agreed. It took them eleven days to drive, cookie jar of brain in tow, from Kansas to Hamilton, Ontario, where they delivered it to McMaster University where it lives today.

More pieces and slices on microscopy slides are held in the Mutter Museum in Philadelphia and in the National Museum of Health and Medicine, Maryland. There will be more sections too, in dusty collections around the world. There was also a slice given to a Japanese mathematician who was making a documentary in search of his hero Einstein. He asked Harvey for a piece and the doctor obliged.

There has never really been anything that stands out, any obvious difference that would point us to a specific brain having belonged to either an academic or a criminal. Maybe they aren't that different underneath. Maybe it's the choices that are made or the circumstances they found themselves in, and not the underlying anatomy, that led them to their life decisions.

We keep looking, in the hope that one day some new technology, a stronger microscope or new understanding will tell us more about the differences between us within our brains.

To that end, the brains of over two hundred Russian scientists are currently kept in the Russian Brain Institute in Moscow. They await the day they can shine once more. Lenin's brain sits there too.

In 1945, ten years before the natural death of Einstein, Mussolini, the fascist dictator, was executed. He was shot and his body was hung up by the feet alongside that of his mistress. Onlookers threw abuse at the dead bodies. What was left of Mussolini's corpse was kept concealed away for years to prevent reprisals, but not all of his remains remained hidden. His brain was taken by the Americans who wanted to study the mind of a dictator. Amongst other things, they tested the brain for syphilis. The test was inconclusive, but testing was not the only agenda here. The brain was also a trophy.

Part of that brain was returned to the Mussolini family in 1966, but not all of it. Some was kept in America and came to light in 2009 when Mussolini's granddaughter made a phone call to the police. She reported that someone was selling vials on the auction site eBay that were supposedly brain segments and blood belonging to Benito Mussolini. Fascist memorabilia are still hot property. The sellers were looking for €15,000 for Il Duce's body parts. If the stories of the corpse being kicked and even shot in the head as it hung on display are true, it can't have been in very good condition. Even in the twenty-first century, body parts were still being stolen as trophies.

Louis Braille's eyes

Our eyes are sensory organs that sit outside of the skull, and as such they are extremely vulnerable. They have some protection around them, as they sit in a bony socket, but they face the outside world to let the light in. They are connected to the brain, which is fully inside the skull's protection, via the optic nerve. A lens directs the light onto the sensor cells at the back of the eye that make up the retina. This nerve gathers the impulses and takes them to the occipital lobe at the rear of the brain. Muscles around the eye control the movements, up, down, left, right, innervated by the cranial nerves. This complicated system can be interrupted at any of these spots. Blindness can be congenital or can be acquired by trauma, infection or autoimmune conditions.

For as long as humans have been telling stories, there has been huge significance associated with the eye and, in particular, the idea of blinding, which has both religious and secular associations. Gods bestowed blindness as either a punishment for sin, or as a gift to permit enlightenment without the burden of sight. In stories throughout the years we have heard of the wise blind man, bestowing a wisdom that has come from the heightening of his other senses. From Greek mythology where we meet Phineus, King of Salmydessus who appears in the account of the voyages of the Argonauts, to Shakespeare's King Lear where the Earl of Gloucester does not truly see what really matters until his eyes are taken, to modern texts and movies, blindness in literature is symbolic. In the 2003 movie *The Matrix Revolutions*, Neo no longer needs his eyes, once he sees the truth of the world around him.

Stories of soldiers being killed by arrows in the eye are plentiful. Both Harold Godwinson at the Battle of Hastings in 1066 and Henry Percy at the Battle of Shrewsbury in 1403 were symbolically said to have died this way. Neither of those stories are believed to be true but are rather morality tales that show fiends dying in a manner that is a fitting punishment for the life they led. Blinding has been used to prevent potential kings or emperors from taking positions of power by their enemies. Gouging of eyes was a medieval torture used together with castration, to prevent the conception of any heir that might challenge the usurpers. In the early twelfth century, when King Henry I's son-in-law blinded a young lad who he was holding as prisoner, the boy's father petitioned the King of England for revenge and the King agreed that his granddaughters should have their eyes put out in forfeit. To do this to his own grand-daughters would completely change their lives and put an end to their prospects. Their mother went after her father Henry with a crossbow.

There are celebratory stories of heroism too: Blind King John of Bohemia charged to his death at the Battle at Crécy in 1346 when he was fifty years old. He had been blind for ten years and still he fought alongside his armies. Born in the eighth century BCE, the Greek poet Homer, author of *The Iliad* and *The Odyssey*, is presumed by some to have been blind. The twentieth century's American author Helen Keller has had statues erected in her honour, and plays written about her life.

Shamefully, blind people were sometimes reduced to second-class citizens, in need of charity or pity. Yet the list of remarkable creatives, musicians, writers and politicians who have been blind is long. There have been blind leaders in all professions throughout history. The blind have been treated with both reverence and derision over the years in all cultures. When Louis Braille was young it was the latter; he complained of disrespect and condescension when he longed to be seen as an equal. Braille was an educator but is remembered best for his wonderful invention that helped the blind to read.

The French Enlightenment era philosopher Denis Diderot wrote a 1749 *letter on the blind for the use of those who can see*, suggesting that touch could be of use as a means for the blind to read. In the very next century, Louis Braille made it happen by creating a tactile system.

Two centuries ago in the village of Coupvray, east of Paris, France, Louis Braille was born in January 1809. This was a time of war for his country; the Emperor Napoleon Bonaparte was on the rampage across Europe. Braille's father was a leather-worker, and one day when Louis was three years old he was playing in his father's workshop. He grabbed a tool called an awl. With a wooden handle and a large needle-like metal point, it is the sort of tool that causes grandmothers to immediately say 'You will have someone's eye out with that'. Louis tried to copy his father at work, attempting to make holes in leather. As he plunged the tool into the piece of leather, he slipped. The sharp point of the awl went straight into the soft fluid-filled eye, piercing it and causing terrible damage. He *did* have someone's eye out with that – his own. The wound did not heal well, and soon became infected. His other eye also succumbed, likely due to sympathetic ophthalmia, a rare bilateral autoimmune response to granulomatous uveitis. Granulomas are collections of white blood cells that clump and form in response to trauma or infection. Here it happened in the uveal tract, the middle layer of the wall of the eye that contains the blood vessel-filled tissue layer, the ciliary body of little muscles and the iris. Sympathetic ophthalmia occurred in response to the injury and claimed young Louis Braille's second eye, and his eyesight with it.

Braille grew up an intelligent and creative young man, keen to learn everything. His father would help him communicate using pins hammered into wood and leather. He went to a local school but of course his reading was limited. He so very much wanted to read like the other children. At ten he was enrolled at The National Institute for the Blind in Paris, where he excelled. There he would learn with the other visually impaired children to use his fingers to try to read embossed lettering. It was hard to do, cumbersome and time-consuming. The books were heavy, and fingering the lettering took a long time to read by touch.

One day, a visitor arrived at the school. Captain Charles Barbier was a soldier who came with an idea he had had on the battlefield. He had developed a system of embossed dots and dashes that could be read by touch in the dark. He knew that lights and noises were potentially deadly as they gave away the position of his soldiers who wanted to sneak about unnoticed, and he could see that a more tactical way of communicating was necessary. His colleagues did not take to the system, partly,

I suspect, as soldiers in cold wet fields at night tend to wear gloves, but also like Braille's experience of embossed lettering, it was still too cumbersome. His superiors were not interested in his crazy ideas either. Isn't that a story as old as time? It occurred to Barbier that his system might be useful for the blind, and so he took his endeavours to the institute in Paris.

Braille was interested and he got to have a play with Barbier's system. He found it to be a good start but, like the embossed lettering of his cumbersome school books, it was difficult to read quickly with the dashes not fitting into the touch of one finger. He wondered if he could improve on the system: getting rid of the dashes, perhaps, and reducing the code to use smaller units. With the very tool that had taken Braille's eyesight, he set to work. He pushed the awl into the card and when he felt the bumps on the other side he knew he had created something very useful. He created a system of six dots arranged into units he called cells, each fitting into the size of a fingerprint. Each cell represented a letter or a code that indicated punctuation, capitals and numbers. Braille's system was much easier, and was faster to read, than anything Barbier had created. Even though he became a teacher at the school, Braille's seniors did not want to entertain this new system and he was told he could not use it. He persisted and published works describing his reading invention. His students liked it – the other teachers refused to use it.

When Braille reached his forties, a tuberculosis infection

caught up with him. He became increasingly incapacitated and had to leave his job as a teacher. In 1852, two days after his forty-third birthday, Braille succumbed to the disease. He could no longer campaign for his reading system or his students, but they continued to push their teachers for its approval. The Braille system was eventually adopted and its use began to spread.

In 1952, one hundred years after his death, Braille's body was exhumed and taken to Paris. There, a procession walked with his coffin through the streets. He was interred at the Pantheon, the mausoleum where the French bury their honoured dead. He lies now with the Curies, Voltaire and Victor Hugo, an acknowledgement of his contribution to the world.

Not all of Louis Braille's remains enjoy an honoured resting place at the Pantheon though. There was something special left behind in his home village. The bones of his hands were removed and buried in a concrete box on top of his original grave. The hands that first touched the damaging awl and the hands that read the first cells of Braille.

Now, there are more than eighty Braille systems based on his nineteenth-century invention. Louis Braille might have lost his eyesight in 1812, but because of that accident he created a system that helped so many people and continues to, to this day. The contribution he has made to the lives of the blind in the last two centuries carries far beyond the borders of France, and his impact cannot be calculated.

Frida Kahlo's eyebrow(s)

Born Magdalena Carmen Frida Kahlo y Calderón, eighteen-year-old Frida Kahlo nearly died in a traffic accident in 1925. She was travelling to school when her bus hit a tram car. A metal handrail was broken off in the collision and when it came into contact with her soft abdomen, it was pushed right through, slicing her organs. But that was not her only injury, bad as it was. As her body collided with the metal frame of the bus, she fractured her pelvis, hips, collarbone, ribs and spine. Yet she lived. What followed was years of pain. She went through thirty operations and at times was laid out flat in her bed for months on end. She took to painting, over her head on a painting stand that her father created for her. Frida Kahlo became an artist.

Kahlo is known for a body of work that included multiple self-portraits, portraits and studies of artefacts from the natural landscape of her home in Mexico. Her subjects were often painted from memory and took on a surreal nature. In her portraits she explored ideas of self, identity, relationships, the human body and death. She has often been labelled a surrealist but she denied it, saying that she simply painted what she felt was true of her condition. Kahlo was also known for her relationship with Diego Rivera, painter of large murals, whom she met when she was fifteen and he thirty-five. Married, divorced and remarried, theirs was a tempestuous relationship with infidelity on both sides. When together they talked much of communist politics and Mexican nationalism, and Frida's art contained representations of Mexican folk iconography. At first Diego Rivera was the most famous of the pair. After her death, Kahlo's fame grew.

It was not just the bus accident that challenged Kahlo's health. She had survived polio when she was younger. It left her with a slight limp, but polio often has other unseen effects that come to bother a survivor later in life. Kahlo originally intended to study the sciences and particularly medicine, but the bus accident put that aside. She underwent so many operations and during her recovery periods it was her painting that took centre stage. She came to express her chronic pain in her paintings. Body parts, particularly transplanted hearts, appeared more than once in her work, but it was her eyebrows that caught people's attention.

Her eyebrows, or rather her one dark full eyebrow that stretched across her forehead, has become an important icon. She never altered her appearance in self-portraits to please others. The long joining eyebrows and the dark wisps of hair on her upper lip were staying. This was her. This was the truth. She kept her strong brows and dark facial hair in defiance. She is now seen as a classic feminist icon and held up as an example of not conforming to the narrow ideals set by other people. If we could teach more young girls to understand the importance of just being the person they want to be, the world and our places within it would be very different.

Eyebrows are a small feature with a couple of useful functions. Physically, the hairs protect the eyes from sweat and other matter dripping down the forehead into them. Secondly they are used to convey feelings in our facial expressions. A raised eyebrow says a lot, even without words. They aren't what we would think of as vital organs, and yet Frida Kahlo's eyebrows found their rightful place in history.

Interestingly, when the makers of Barbie dolls released a Frida Kahlo doll in 2018, as a 'celebration of strong women', they completely missed the point. There's no doubt this feminist icon was worthy of her own doll, but they split her single eyebrow into two and there was no obvious facial hair elsewhere.

Barbie minimised the very thing that made Kahlo's appearance stand out. Anyone buying a Frida Kahlo Barbie doll needs to buy a Sharpie pen with it.

Later in life Kahlo developed gangrene in her leg and needed amputation.

Since her death, we've become more and more enamoured with Frida Kahlo and what she and her eyebrows represent, even if dollmakers have a way to go in representing her.

Tycho Brahe's nose

It's a classic story. Two young men, under the influence of drink, were unable to settle an argument over who was better at mathematics. OK, maybe the mathematics part isn't classic, but one of the young men was the Danish astronomer Tycho Brahe. 'I demand satisfaction,' his adversary shouted as he hit Brahe across the face with a challenging glove. He got his satisfaction. The duel was fought with swords and Tycho Brahe's face was never the same again.

To say Tycho Brahe was just a Danish astronomer is a bit of an understatement. He redrew the solar system and was the first to understand that supernovae were not comets. He brought science to astrology, the influential study of the stars that was thought to rule our fate and lives. Brahe questioned the prevailing Ptolemaic theories, and he did it all with the naked eye (albeit without a nose). Tycho Brahe knew, as many a long-term sufferer of syphilis can testify, one can live without a nose.

Born in 1546 to a powerful, noble family who were friends of the King of Denmark, Brahe was the eldest of twelve siblings. Eight of them made it to adulthood, an unusually high rate of survival. Brahe's twin sibling died when they were young, and not long later, Tycho was effectively kidnapped by his uncle. He was raised by him from the age of two as if he were his own. Luckily for Tycho his uncle was a rich man and young Brahe wanted for nothing. He was well-educated and at first he attended the University of Copenhagen to study law. There are conflicting stories of Brahe's decision to switch to the study of astronomy. One claims that at about fourteen years old, he witnessed a solar eclipse that had been predicted and he was so amazed that such a prediction could be made, that he went about figuring it all out for himself. The other story is that he witnessed the solar eclipse in August 1560, but it had come a day later than had been predicted. He then decided that he could do a better job. Either way the teenage Brahe was taken with the study of the stars and fortunately he was quite good at the mathematics that went with it.

The prevailing Ptolemaic theory was that the solar system was geocentric, that everything in space revolved around the earth.

Brahe questioned that, and although Copernicus had already identified that the solar system was heliocentric, Brahe built on the ideas. Impressively, all of this work was achieved long before the introduction of the telescope. The current theory was that everything beyond the moon was fixed in place. Brahe had noticed a new shine in the sky, in the Cassiopeia constellation. This was a supernova and it changed all thoughts of immovable celestial beings. Brahe decided that, whilst all the other planets orbited the sun, the earth was fixed and the sun went around the earth, so he kindly left some discoveries to others.

Due to these huge findings and his writings, his fame grew. Brahe was patronised by Frederick II, King of Denmark and Norway, and he worked from an impressive castle observatory built for him on an island granted to him by the King. There, he continued to drink and brawl. He treated the locals very badly and he annoyed people in high places when he married a commoner. He led quite the life, so it did not matter so much that his nose was missing. For the rest of his life he glued a prosthetic brass nose to his face. At least he had somewhere to rest his glasses.

As well as a place to rest your spectacles, the nose does have a useful physiological function: with the nasal intake of every breath, air swirls past receptors in the lining of the nose that sample molecules as they pass through. From the receptors, signals travel along the nerves to the brain and there, in the grey matter, a decision is made. Is this smell good or bad? Is this thing we can smell fine to be around, or should we withdraw from it? As the air enters the nose it is warmed in the passages so that it does not hit the lungs cold. The warm environment in the lungs needs to be maintained to be optimal for cell functions. The air is also filtered in the nasal passages, the nose being a gatekeeper to the potentially dangerous particles flying around us.

Smells can be so provocative of memories. They can make us recall wonderfully happy events but also the dreadful experiences we might prefer to forget. There is a physiological explanation. Smell is not controlled by the hypothalamus, instead smells are handled in the olfactory cortex, also the place where memories are made. This may seem a strange quirk of the brain but in fact it's important to make memories of smells, as doing so could save a life. The sweet smells coming from the kitchen at dinnertime can signal good things ahead. Rotting flesh is best avoided as it brings risk of infection, and so we perceive the smell as deadly. It's not all plain smelling though. There are potentially dangerous substances that can lead us astray. Who doesn't love the smell of freshly cut grass? That's the smell that deadly nerve agents classically give off. Sorry if that has ruined the summer for you.

Losing one's sense of smell can happen without the loss of the nose itself. Smell can be knocked off temporarily by an infection of flu or perhaps Covid. It can be knocked out permanently by a head injury affecting the area of the brain where the signal is interpreted. In none of these incidents is the nose itself lost. Losing the nose, though, can be an outward sign of something sinister.

Syphilis was known as the great imitator, having so many manifestations. But one sure-fire way to recognise a sufferer of the disease was when the bridge of the nose collapsed and the nose disintegrated. Losing one's nose to this disease would likely come after years of infection and with a long list of other ailments and complaints, pains and sufferings. As there was no effective treatment for syphilis before the twentieth century, so many who contracted the disease went on to suffer from the later stages of syphilis infection in this way, losing their noses. In the early 1700s, there were enough noseless sufferers of syphilis that they could hold their own club, meeting monthly for dinner in London's Covent Garden. They met in a pub and sat down to a meal of pork from pigs cooked with the snouts cut off. They would wear false noses often made from leather or, like in Brahe's case, a prosthetic nose of something more solid. A skin graft was sometimes attempted with skin taken from the sufferers own arm to cover the holes left in the face. Brahe told people that his shiny nose was made of gold. In 2012 an exhumation of his remains discovered that his nose was made of brass. If

I were his family member, I'm not sure I would have buried him in his Sunday best though.

Whilst Tycho Brahe is remembered for having lost his nose, other notable noses have found their place in history for sticking about, or even sticking out.

Rudolf I, the first Habsburg king of Germany, had a nose so big that artists refused to paint it in its original dimensions. Perhaps they were worried about using too much paint or, more likely, the potential reprisals. Finding an image online of how the portrait artists represented Rudolf I's nose can be challenging, given the other Rudolph and his very famous red nose.

That's just one of the many gigantic noses from history. Hercule-Savinien Cyrano de Bergerac was a real man who the synonymous play was based on. He might have been an over-the-top character in the play, but in reality he was a seventeenth-century swashbuckling poet who sported a huge nose. He was fond of duelling and his nose must have been a tasty-looking target.

Standing out from the crowd for his huge nose is probably not what outlaw George Parrott would have wanted. He was a nineteenth-century cattle rustler who happened to have an enormous nose. His nickname, like his nose, went before him wherever he went. In 1878, Big Nose George and his gang of highwaymen were planning to rob a train and so tampered with the railway tracks to bring it to a halt. The tampering was spotted though, and the law were called in. The gang went on the run, followed by a Sheriff Widdowfield and a detective named Tip Vincent. When the pair of lawmen came across a campfire they stopped to investigate. The fire had recently been burning and they realised the gang must be close. In fact, they were too close. Gunfire rang out and Vincent was shot. The sheriff ran in to help Vincent and he, too, was shot. The railroad put a bounty on the head of Big Nose George Parrot.

Two years later, George was caught and sentenced to hang. A week or so before the date of his execution, he tried to escape but the town were having none of it and Big Nose was lynched. Strung up on a telegraph pole.

Nobody came forward to claim the criminal's corpse and so a couple of doctors called Osborne and Maghee stepped in. They wanted to examine his brain, to see if in the brain of a criminal there were clues to his criminality, but they did not stop there. Osborne took a knife to the body, skinned the cadaver and sent the skin off to a tannery to be made into a pair of shoes. Osborne later became the first Democratic Governor of Wyoming and was rumoured to have worn the shoes to his inaugural ball.

The top of his skull, which had been sawn off to get at the brain, was handed to Osborne's young assistant, a girl named Lillian Heath. Lillian, with all that young experience, went on to become Wyoming's first female doctor and she used the skull cap as an ashtray and a doorstop at her office.

The rest of Big Nose's body was dissected and examined, and kept in a whiskey barrel full of salt solution as the doctors took it apart. When they were done, the barrel and the remains were buried in the back yard. They lay there, unseen, for a hundred years, until 1950 when, during building work, the barrel was found. It was opened and the skeleton was identified using the remaining sawn-off scalp which was still in the possession of an aged Dr Lillian Heath. The remains were identified as those of Big Nose George Parrot. Sadly his big nose, which was made of soft tissue, was no longer there. It was either dissected away by intrigued doctors post-mortem, or it decayed with the other soft tissue over time. There were no signs on the skull to say exactly *how* big Big George's nose was. Nobody knows what happened to his nose.

Marie Antoinette's teeth

'Let them eat cake,' she said, because it's easier to chew than a steak. That's what Marie Antoinette really meant when she said those famous words, isn't it? She was one of France's most famous queens, ruling with her husband King Louis XVI in the eighteenth century. This line about cake (or maybe brioche) is attributed to her, but whether she said it at all about the starving peasants is contentious. There was no shortage of enemies lining up to say bad things about Marie and put words into her mouth. After all, words were not the only thing they put in her mouth.

The upper classes were keen to unite the ruling house of France, the Bourbons, with that of the Holy Roman Empire, and chose Marie from childhood for the task. Marie was moulded – quite literally, as we shall see – from a young age into the queen she became. She was an Archduchess of Austria, the daughter of Maria Theresa, the Holy Roman Empress. It was a wonderful political match for the future king of France, to address a rift between France and Austria. There was a slight problem though – by courtly standards, the young Marie Antoinette was a bit of a mess. When the envoy, the Duc du Choiseul, travelled to meet the young girl that could one day be Queen, he was far from impressed with her. There was a lot of work to be done if the young girl with creased clothes and messy hair was going to be married to the Dauphin, the heir to the throne of France. Marie Antoinette had Habsburg blood and with that came Habsburg genes. Habsburg genes famously made for interesting jaw lines and wonky teeth.

In the French court, fashion and looks were everything and Marie's teeth were sadly thought to fall short of the standards required of a future queen. Before she could marry the Dauphin, Marie Antoinette had to sort out her smile. The answer was one of the first uses of braces.

The job of fixing Marie's teeth was given to a celebrity dentist by the name of Pierre Laveran. He used a device that had been invented by another well-known dentist and was known as Fauchard's Bandeau. Even the name sounds painful enough. The metal horseshoe-shaped arch was placed inside her mouth. It had a row of tiny holes through which gold wires were pushed and then tightly wound around the teeth. It reshaped the arch of Marie's mouth and straightened her teeth. Over the next few months the wires would be tightened and tightened again, until Marie Antoinette had a smile fit for a queen. It would have been painful and relentless. When she was fourteen years old, with much straighter teeth, she met her future husband. It is a surprise after all that pain that she was able to smile, though being betrothed to the Dauphin of France probably helped.

Fauchard's Bandeau had done the job. Pierre Fauchard (1679–1761) was a French physician who is often credited with being the father of modern dentistry. His interest in the subject arose when he worked as a ship's surgeon in the Navy. There he saw first-hand the ravages of scurvy. When there is a deficiency of vitamin C, collagen is destroyed, leaving gums to rot and teeth to fall out. The disease was commonplace amongst sailors spending so long at sea away from common sources of vitamin C in their regular diets. It is understandable that a Navy surgeon wanted to solve that particular gruesome problem. He described no less than 103 different types of oral diseases and provided suggestions on how to treat them all. Fauchard studied the work of watchmakers, jewellers and barbers and adapted their skills and tools to work with teeth and gums. He devoted a whole chapter of his famous book on dentistry to the straightening of teeth.

In 1793 when the French rebelled against the ruling classes, Marie Antoinette's teeth were as straight as the sharp blade of the guillotine that chopped her head clean off, perhaps rolling into the basket with one last perfect smile.

Whilst Marie Antoinette was losing her head in France, across the Atlantic George Washington was serving as the first president of the newly formed United States. He too had a problem with his teeth but needed something more than gold-wired braces.

As he took charge of Revolutionary War armies, famously crossed the Delaware and was eventually sworn in as the first president, he suffered constant pain in his mouth. His gum problems started when he was young and the pain left him unable to eat anything but a simple diet of cornbread and soup. He would brush and take care of his teeth but it was not enough to save them. He was in his early twenties when his first adult tooth came out and by the time of his first inauguration when he was fifty-seven, he had only one tooth of his own left. Whatever his initial pathology, it is thought that mercury treatments contributed to his failing teeth. Mercury was used as a treatment for many ailments, but it did far more harm than good. Washington needed dentures and he was fortunate enough that he could afford to have a few pairs.

There were two dentists of note who looked after the

president's gnashers. Jean Pierre Le Mayeur was a French dentist who initially served the British until he became infuriated by how they treated the French. He packed his tools and marched across enemy lines to look after the teeth of the Americans. He was welcomed and called upon by Washington to take a look in his sore mouth.

Another dentist by the name of Greenwood was responsible for Washington's famous dentures. Contrary to popular belief, they were not made of wood. They looked that way because of staining and discolouration but they were made of ivory from elephants and rhino as well as lead, embedded with teeth from animals and humans. It's known that slaves at his home at Mount Vernon had been paid for their teeth but it is not known if the dentures that are on display today used any of their particular teeth. Wherever those teeth ended up, it can be said that the pulling of them, in a time before anaesthetic, would have been traumatic and painful. It's not hard to imagine that, even though payment was made, they probably weren't a hundred per cent willing participants in the affair.

The president's dentures were ill-fitting and caused as much pain as his diseased gums. They would push his lips out and they reshaped his cheeks after years of use. Washington was embarrassed by his dental disease and so he mostly kept his mouth shut. If he needed to speak, he kept his speeches short, worried that the spring-loaded rows of teeth might shoot out from his mouth when he addressed a crowd. Portraits of Washington show the facial distortion caused by the dentures. Gilbert Stuart's portrait in particular, used on the US one-dollar bill, shows an obvious issue around George's jaw. Perhaps Washington's famously short temper was due to his constant pain and having a mouth full of lead and other people's teeth. That would make anyone a little grumpy.

The need for dentures was so high that teeth were stripped from the mouths of criminals as they were hanged on the gallows. They were plied from the mouths of the dead after body-snatchers dug them from their graves, and they were cut from the jaws of soldiers who lay dead, killed on the battlefield. So many teeth were collected from the dead of the Napoleonic battlefields that in Britain and Europe in the nineteenth century they became known as Waterloo teeth. Better to take the teeth of young healthy soldiers who had died in their prime before disease had a chance to set in. Nobody wants to make their dentures out of someone else's rotting teeth.

When John Greenwood removed Washington's last tooth he kept hold of it and had it mounted in a locket that he kept on his watch. His toothy memento, along with some of the dentures and his impressive array of tools, can be seen at the New York Academy of Medicine. Another set of the president's dentures can be seen at Mount Vernon, vital organs

Washington's home, which is now a museum. They don't look like anything I'd like to put in my mouth, but then it must have been nice to eat something more than soft and squishy cornbread occasionally.

Charles II of Spain's jaw

In 1661 a boy was born to the Habsburg dynasty who would be king. It was a miracle he had survived the birth at all. In whispers he was called El Hechizado – the bewitched. King Charles II of Spain was born with the classic Hapsburg jaw. It's no wonder. Well, his jaw might have been a wonder, but it is no surprise that he had it. The King was the product of generation upon generation of inbreeding. The result was a boy born with severe physical and mental problems. What Charles went through was awful and it was no fault of his own. Charles, or Carlos as he was known to his Spanish friends, was born to a dynasty who liked to keep everything in the family. Land, property, wealth and power were all kept close to home by marrying the nearest of relatives. The Habsburgs rose to power in Austria, Germany and the Holy Roman Empire, from Portugal to Transylvania, and they were not sharing it with anyone.

Two centuries before, in 1496, Philip I, the Archduke of Austria, a good-looking chap known as Philip the Handsome, married Joanna of Castile, thus bringing Spanish royalty into his family. This was excellent news, the new blood a good thing, but over the generations it all went awry. For years, uncles procreated with nieces and cousins with cousins. In marrying so close the family risked genetic disaster. Genetic homozygosity, the inheritance of the same genes, occurs when family members have children together. Mistakes or mutations in the genetic code are often not noticed when strangers mix genes up, except in rare cases when recessive genes come together. If bloodlines are too close, those recessive genes meet more often and the result is an anomaly in the phenotype – the outwardly observable traits of our DNA, and that's what Charles suffered with. In the Habsburg family there were a few classic physical complications. Lower lips were obviously bulbous, noses could be excessively long, but the most obvious was the characteristic protruding jaw.

Charles's jaw was so big and deformed it left him with a huge underbite. His teeth at the top did not meet those at the bottom. He could not chew his food and as a result gulped down big chunks of his fare causing him to dribble. A British envoy called Alexander Stanhope was a minister in Madrid in the late seventeenth century. He wrote a letter home to the Duke of Shrewsbury and in it he described the King's features. 'He has a ravenous stomach, and swallows all he eats whole for his nether jaw stands so much out, that his two rows of teeth cannot meet.'

Upon his father Philip's death, Charles became king. He was only four years old and so his mother acted as regent. Charles's mother was Philip's niece, herself the daughter of kissing cousins. Their boy did not speak before he was six and did not walk before he was eight. It was not even thought to be worthwhile trying to educate him. When he reached the age when other young kings take over the reins, it was simply not possible. Charles's mother continued to be influential in court and Charles remained infantile. He only worked for a quarter of an hour each day when he would sign papers placed in front of him. Otherwise he was at play, frivolously running about from room to room like a child, or he would be out hunting. Decisions were made for him. Martin A. S. Hume in his 1905 book *Spain: Its Greatness and Decay (1497–1788)*, described a time of semi-anarchy where officials were left to deal with social problems and would often pass decisions off to others.

El Hechizado was married twice but unsurprisingly he did not have any children. That was not thought to be the fault of either of his wives. His first wife, Marie Louise of Orleans, died in 1689 of likely appendicitis, though of course poisoning was blamed. Was someone after revenge for her not giving the kingdom an heir? Charles was immediately married again, but his new queen Maria Anna of Neuburg was not to have his child either. With no direct heir, this was the beginning of the end for the ruling house. Charles was to be the last Habsburg King of Spain and as his health failed the vultures were circling. Different parties fought for control over the King. Decisions were made for him. He signed a decree that Philip of Anjou should be his heir, then he was later persuaded that it should be the Austrian Archduke Charles, though he didn't sign anything to that effect. The dying king was talked into changing his mind so often that there was much confusion. Upon his death, Philip of Anjou was pronounced King, but Austria had other ideas and the Grand Alliance of European nations were threatened by the power shifts created by such vast territories going to France. Charles II's death sparked the War of the Spanish Succession, and many, many more deaths were to follow.

It was not known for a long time whether the Habsburg jaw was merely a family characteristic or if it was the result of family inbreeding. Geneticist Roman Vilas and his team at Spain's University of Santiago de Compostela spoke to ten maxillofacial surgeons who each studied sixty-six different portraits of members of the Habsburg family. They identified the features of mandibular prognathism (protrusion) and maxillary deficiency. A deficiency with the maxilla, the bone that sits above the jaw, is linked to a protrusion in the jaw. They scored each family member according to the degree of deformity in these areas. The team also looked at the family tree to identify how closely each family member was related. What they found was that the more prominent the jaw, the higher the inbreeding coefficient (the more likely they were to be related). There was the answer. The characteristic Habsburg jaw was a genetic product of the inbreeding. Those with a bigger jaw were more likely to be born to closely related parents. Incredibly, Charles II's coefficient was even higher than had his parents been siblings, so frequent had the inbreeding been over generations before him. It was remarkable that Charles made it to adulthood, but there were many other Habsburg children who did not because their genetic complications were so bad.

The family's inbreeding also bred an infant mortality of 50 per cent. That was much higher than the average for the time. Charles was one of ten children and he was the only boy to survive. That he did was a miracle, but he was unwell throughout his whole life. He died when he was thirty-eight years old.

The autopsy of Charles II of Spain was brutal in its reporting. The physician found Charles's heart to be the size of a peppercorn, with not one drop of circulating blood. He had one blackened, shriveled testicle. His guts were rotten and his brain was full of water. Embellished, of course, but chilling none the less. This was a warning to all. Families need new blood or there will be extreme suffering for those who did not ask to be born that way.

The familiar portrait of Charles II of Spain featuring the trademark jaw, painted by Juan Carreño de Miranda, was no doubt also embellished, but this time with a more positive spin to flatter the King. Goodness knows how that jaw really looked.

Charles was not the only one to sport the famous family jaw, but he suffered the most with it. Following his death, the War of the Spanish Succession lasted for fifteen years. For a decade and a half, war raged across Europe because Charles II of Spain was too broken to provide an heir and save the dynasty that broke him.

Tatiana Pronchishcheva's gums

When Russia's Peter the Great looked over his maps early in the eighteenth century, he saw huge empty spaces. 'Here be dragons' did not cut it with the Tsar, and so he funded an ambitious project to colour in the map. His plans were to chart the Arctic coast of Siberia and parts of the North American coastline. Or, at least, have somebody else do it for him. The project would require lots of sturdy ships and lots of equally sturdy sailors. It was unusual for women to join their husbands on expeditions, but not wanting to be left home alone, Tatiana Pronchishcheva did exactly that, making her the first female polar explorer. Together the Pronchishchevs charted the coastlines and recorded the natural history that they found. It was quite the adventure.

In 1735 they travelled down the Lena River from Yakutsk, but cold, dark, icy nights were on the way and so they had to stop for the winter at the mouth of the Olenyok River. Sadly, the ship's crew did not fare well when overwintering in a place where it is frozen for over eight months of the year. They waited it out. Many fell ill and died, including the Pronchishchevs, who are believed to have died the agonising death of scurvy. Scurvy is a nasty illness caused by a deficiency of vitamin C in the diet, that has the potential to destroy navies and end expeditions.

In 1999, members of the Russian Adventure Club met in a McDonald's restaurant in Moscow. Fuelled by Big Macs and fries and not a lot of vegetables (with no record of whether or not the milkshake machine was working), they devised a plan to go and find the bodies of the Pronchishchevs. There were records of the remains of an eighteenth-century settlement in the Bulunsky region of the Sakha Republic (Yakutia) in Northern Siberian Russia, and local legends told of a grave with an old cross that marked the resting place of the married explorers. The group of adventurers set off in search of the bodies and when they found the grave site, out came the shovels.

In the grave, they found two skeletons lying next to each other. A male and a female. The adventurers described both skeletons as being of European build, from the western region of Russia. Spectrum analysis of the bones confirmed that they were not the remains of people local to this region. In their graves were artefacts and clothing also suggestive that the remains were of western visitors to this part of the vast land area that is Russia. Whoever they were, this couple died far from home.

Vasily Pronchishchev was a Lieutenant in the Navy. From the age of fourteen he studied at the Moscow School of Mathematics and Navigation. In 1721 Tatiana's family moved to the port city of Kronstadt where she met Vasily who was stationed on the north coast. He served with Peter the Great's fleet and later became a commander of the detachment of the Tsar's Great Northern Expedition. Under the leadership of Vitus Bering, the Captain commander and navigator, they mapped the Russian territories. Vasily and Tatiana married in 1733 and when he was given command of a vessel he took his new wife with him. It's not even known if he sought permission. It was unusual and so Tatiana Pronchishcheva is thought to be the first woman to join such an expedition.

When the investigators examined the skulls, they found no signs of scurvy in Vasily's teeth or jaw. There was an interesting injury, though, to his leg. Vasily had been hit by a sharp projectile from behind. It had punctured his skin and made a hole in his tibia. This injury had occurred a couple of weeks before death as the start of the healing process could be seen. Vasily's death was beginning to look a little less like scurvy than the reports told us.

A break in the bone like this could have killed Vasily. The wound might easily have developed a lethal infection. A fat embolus might have broken free from the fracture and moved through the blood stream to become embedded in a small artery, causing a quicker death. There is no way of knowing from the bones that lay below Russian soil what killed him. His soft tissue would have been useful, but he had laid in the ground for nearly two hundred years. Given the evidence of a broken leg, that scurvy killed Vasily Pronchishchev is debatable, but it could well have been that he also had scurvy when he was injured. We are often quick to assign only one diagnosis when more often than not they come as a package. Wounds are slow to heal, if at all, when the body is deficient in vitamin C, so an injury to a scurvy sufferer could be disastrous.

As for his wife, not much was written of the first female arctic explorer. When occasionally she gets a mention in the records of the navigator's logbook she is 'the wife of Mr Lieutenant'. Her name was Tatiana but even that detail was lost until 1982, when researchers found it written in old papers that had been sitting in a Russian archive. Before that it was taken for granted that her name was probably Maria. The name stuck, as did the assumption that the couple died of scurvy. Some ideas are hard to shake off once they have been copied and pasted around the internet a few dozen times, so you will still commonly see both the incorrect name and a possibly incorrect diagnosis.

As for Tatiana's remains, there were no obvious signs that she suffered a trauma like her husband. Her skull showed no overwhelming evidence of chronic disease either, though two teeth were missing. The soft tissues that degrade with the disease of scurvy had long since returned to the earth and so bones were the only evidence.

So why would it be that the Pronchishchevs were recorded as having died of scurvy if they hadn't? There are lots of reasons why deaths are recorded the way they are. Assumptions are sometimes made, or worse, the records are written to benefit those left behind, for ill gain. However, there is always the danger that we might be trying to see something that's just not there. The most obvious reason for their deaths is scurvy, as so many of the expedition crews were recorded as dying from the disease. Common things are common, as any diagnostician will tell us. Tatiana's missing teeth might indeed have fallen out when her gums rotted away with scurvy.

Scurvy occurs when there is a prolonged lack of vitamin C. A vitamin is an organic compound required for the body's function, but our bodies are not able to produce it themselves. Therefore, vitamins are an essential part of our diet. Humans, primates, guinea pigs and fruit bats, randomly, are the only ones amongst us that do not synthesize vitamin C on our own. We must get it from the foods that we eat. Vitamin C, also known as ascorbic acid, is so vital because it is used in the production of collagen, a structural protein that makes up the connective tissues. Collagen is found in joints, skin, vasculature, organs, everywhere. We need a lot of collagen, and the vitamin C used to make it has a high turnover. When collagen breaks down, if it can't be replaced, the result is ulcers, bleeding and rotting. People with scurvy rot whilst they are still alive.

Scurvy was such a problem three hundred years ago that the owners of the ships presumed that they would lose 50 per cent of their crews to it, and accounted and recruited accordingly. It's no surprise the navies needed men so badly they would press gang the vulnerable into service, but sailors were not alone in suffering from scurvy. It was not a disease that was caused by the sea air. It was seen elsewhere, amongst the poor and the displaced. Sailors were a captive audience, though. Stuck on ships for months on end, with only the provisions they had aboard unless they made landfall. They were so vulnerable.

The sailors' diets of preserved meats with oats, gruel and sailors' biscuits just didn't cut the mustard. Come to think of it, mustard would have been handy, plenty of vitamin C there.

There was some understanding of scurvy before the Pron-chishchevs' Russian expedition lost so many crew members. In 1535 it was reported by the French explorer, Jacques Cartier, that natives had helped him when he got stuck in the frozen St Lawrence River. They had provided tea made of leaves and bark, and none of his crew developed scurvy. It was believed by some captains that sailors were saved from scurvy when they had fresher fruit and vegetables aboard ship. In 1734 Johannes Bachstrom, a Dutch physician, first used the term antiscorbutic (meaning against scurvy) when describing fresh vegetables. So, it was understood that there was a preventative measure that could be taken against the disease and yet there was no agreed standard solution.

In 1747 a Scottish Navy surgeon, James Lind, went looking for answers. He knew that some foods could be antiscorbutic but not which ones exactly, or how best they should be prepared or stored. He carried out experiments on his shipmates aboard the naval ship HMS *Salisbury*. He is credited with carrying out one of the first controlled experiments after subjecting twelve patients to six different treatments. A small group, but it was a start. He concluded that fresh fruit might help. He wrote his *Treatise on Scurvy* in 1753, which has been criticised since but did contribute to the knowledge of the day. As is still the case with medical experiments and newly acquired knowledge, it took a long time to get through to the decision-makers. A stumbling block was getting fresh fruit juice aboard ship, as boiling it to make potions destroyed the vitamin C.

In the 1730s, the Pronchishchevs and their crews did not

have the answers either. If Tatiana had been suffering from scurvy, she would have at first felt slow, lethargic, her legs heavy. Her joints would have ached and her skin would have bruised at the slightest touch. As her gums broke down, they would feel spongy and would bleed. At first the gums might swell so badly that they covered over the teeth, needing the ship's surgeon to cut the bleeding gums away with a knife. Then the bloodied rotting gums would recede so far that teeth would dislodge and fall out. The breath of a scurvy sufferer would stink. Mucous membranes, the lining of the mouth, and the linings of the blood vessels are made with collagen and so that could mean severe nose bleeds. Limbs would swell. Sufferers were said to have taken knives to the rotten dead flesh on their own bodies, cutting it away. Old wounds would break open, and if a scurvy sailor had broken a bone in the past then they might expect the callous of healed bones to break again. Treatments centred on balancing the humours by bleeding, purging and using emetics, causing further pain and damage to the body. The four humours still being the predominant theory in medicine at the time, it was thought that an imbalance of humours blocking the pores and sweat glands was the cause of scurvy.

Lind published his suggestions in 1753, but it was not until 1795 that another Navy surgeon, Dr Gilbert Blane, convinced the Navy to bring lemon juice along, issuing it to sailors aboard ship. It changed a lot for the sailors, but it didn't change everything. Scurvy was still found amongst expeditions, soldiers in camps and the poor. Being the result of a vitamin deficiency, scurvy cannot be eradicated like an infection such as smallpox. As long as there are wars, poverty and famines, there will be scurvy.

British sailors carried limes to prevent scurvy, which earned them the nickname 'Limeys', but vitamin C can come from other fruits and vegetables and leafy greens, even onions. That vitamin C comes naturally in packages with sugary or starchy food is telling. Innuits and committed carnivores have shown that extra dietary vitamin C is only needed when eating carbohydrates. This appears to be because the vitamin has a similar structure to glucose, competing with it for receptors on our cells. If you're going to just eat meat and avoid scurvy, the meat needs to be fresh. That was not an option when spending months aboard ship. Lemons and limes were far easier to take on voyages than fresh meat, though even then some thought the fruit unnecessary and too expensive.

Tatiana Pronchishcheva was young when she died, only twenty-six years old. Imagine what she could have achieved and where she could have travelled if scurvy had not been ravaging seafarers across the world in the eighteenth century.

Robert Jenkins' ear

In the year 1731, sea captain Robert Jenkins stepped onto the gangplank and climbed aboard his British merchant ship *Rebecca*. He ordered his crew to set sail and they left Jamaica, bound for London, England. Also on board the ship was the much-loved commodity of sugar. In this case, the ship's precious cargo was true to the records, but often the ships in these seas carried goods not honestly matching the ship's inventories. Contraband was in high demand and the pull of the black market was strong.

In prior years, Queen Anne and her Hanoverian successors had signed a series of treaties regarding trade in the Caribbean. In 1715 at the end of the War of the Spanish Succession, the British were awarded the Asiento as part of the Treaty of

Utrecht. The Asiento was a monopoly granted by Spain to the British to supply slaves to Spain's American colonies, 5,000 of them a year, which the British passed to the South Sea Company. The Spanish had no influence in West Africa from where the slaves were being taken, but did rely on slaves for their exploits in the New World, and so they traded with those who dealt in enslaved people, like the British. As well as the trade of slaves, the British were allowed two ships per year selling 500 tons each of goods in Porto Bello in present-day Panama and Ceracruz in present-day Mexico. The French were earning well there, saturating the ports with French goods, and the British wanted access too. The Spanish could stop and search vessels in the Caribbean to ensure that the British were sticking to the agreements. The day that the *Rebecca* left Jamaica was to be one of those days.

The coastguard ship *Isabella* demanded *Rebecca* be stopped. The British ship was boarded and searches were made. Nothing untoward was found but the Spanish captain, the privateer Juan de León Fandiño, was not happy as he was convinced there was contraband onboard. The ship was searched again and a crew member was beaten. Captain Jenkins looked on but gave nothing away. Then they turned on him. He had a rope thrown round his neck and he was hanged, though not to the point of killing him. Jenkins still had nothing to reveal to the privateer. He claimed to have nothing hidden on board besides a small amount of money that would see them safely back to England. The Spanish were not appeased. Jenkins was held down and his ear was cut at and ripped off, with a warning to the King that should he think of doing a spot of smuggling, he would suffer the same fate. Jenkins was left mutilated, bloodied and angry.

The ear comprises of two main parts. The outer ear, known as the pinna, is the part that we can see, and luckily for Jenkins, hearing is not totally lost if the pinna is damaged or removed. The pinna is made of cartilage but has, as anyone knows who has ever cut their ear, a reasonable blood supply. Inside the ear, the tube, the hairs, the wax, the drum and the tiniest of all the body's bones can still function even if the pinna is lost. The pinna channels sounds into the inner ear, so without it, hearing is more difficult. Any scar or scab forming in the area could impact those sounds getting through, but to lose one's outer ear is not the end of the world for most people. Unless it starts a war.

As in any scenario when the skin is compromised there is a chance that infection might follow, or worse, that it might result in sepsis, a systemic response to the local infection. Jenkins survived the attack though and he sailed back to England with his ear in a bottle.

When Jenkins and his dishevelled crew arrived back in London, there was a ripple of interest in the sea captain's story. He presented his pickled ear to the King in protest of his treatment under the agreement with the Spanish. It all went quiet, however, and not because Jenkins couldn't hear anything . . . but because politics is politics. When it was convenient, Jenkins' ear was heard from once again. Years later when the Tories were looking to oust a long-standing Whig government under Walpole, they looked for a fight. The many stories reaching British shores telling of outrageous Spanish behaviour were raised and no doubt embellished, but Jenkins' incident was not an isolated one.

The stories suggested that Jenkins and other captains in the Caribbean risked losing their cargo, their ships, and maybe even their lives. One story told of a Dutch captain who was forced to eat his own hand after the Spanish coastguard chopped it off. The torturers kindly boiled it for him first. Jenkins was called upon once more and paraded with his severed ear in a bottle in front of Parliament. Justice was demanded, but increased trade in the Caribbean was an even better objective. Jenkins and his missing ear had now become political capital. In a sketch of the time Walpole is seen swooning at the sight of the severed body part. The British wanted more from the West Indies and they used Jenkins' ear as justification, so war became inevitable.

Thomas Carlyle named the conflict 'The War for Jenkins' Ear' when he wrote about it over a hundred years later. The name stuck because we all love a good hook. The Spanish refer to the war as the Guerra del Asiento, the Asiento being the trade agreement between Spain and other nations, particularly for slaves. The war had a rocky start for Spain but in the end they defended their positions. The British attempt to gain territory and further their trade opportunities was thwarted. After the War for Jenkins' Ear, the British lost the Asiento in 1750 with the Treaty of Madrid.

It has been rather sensationally claimed by one commentator that Jenkins' ear is the most famous ear in history and yet Gaudi's book on the subject called it the 'forgotten war'. As for Jenkins' ear being the most famous severed ear in history? I'm not so sure. There's another, more recent severed ear that springs to mind.

In December of 1888, surrounded by paints and canvases, the artist Vincent Van Gogh was living in a house in Arles with his friend and fellow artist Gauguin. Relationships though, were not among Vincent's strong points. The artists were not getting on and an argument led to Gauguin threatening to return to Paris. On 23 December he walked out after Van Gogh handed him a cutting from the newspaper. It read 'the murderer took flight'. Van Gogh had become aggressive, upset that Gauguin threatened to leave. Gauguin was a bit annoyed and he took himself off to a hotel. Later that night Van Gogh took a razor blade to his left ear, cutting into his own flesh. He wrapped up the severed ear, now a lump of bloodied cartilage and skin, and he took it to a brothel he frequented. He handed it over, asking for it to be given to a girl called Rachel, with the request that they 'guard the object carefully'.

The next morning a policeman found Van Gogh lying in bloodstained sheets. When he did not respond, the policeman assumed the artist was dead. You would, wouldn't you? There was blood all over the house. Despite the dramatic blood loss he had survived. He was admitted to hospital and when he woke he denied any recollection of the events. He could not explain why he had mutilated his own ear. He was diagnosed with acute mania and generalised delirium. He later voluntarily entered an asylum at Saint-Rémy-du-Provence to recover.

Many have tried to explain Van Gogh cutting off his own ear, but there is no one conclusive theory. One school of thought cites a popular book of the time that involved an ear being cut off as punishment. Perhaps Van Gogh was inspired by the story and so popular culture must have been a bad influence. Another said there might be a connection to Jack the Ripper, who cut off the ear of one of his victims, which is shockingly tenuous. Others make the connection to Peter cutting off the ear of Malchus after Judas betrayed Christ.

Cropping, or the cutting-off of ears, has been used as a punishment throughout history and the world. Others had their ears nailed to a pillory post and had to lose the ear to free themselves. In 1538 in England, a lad named Thomas

Barrie reputedly died from shock after his ear was cropped. His crime was spreading rumours of the death of Henry VIII, serious indeed. In the same period one could face cropping for the crime of vagrancy, and if you ran out of ears and yet were convicted again, you could hang. In seventeenth-century England, ears were cut off those publishing religious views contrary to the establishment. Cropping comes up in ancient Assyrian law and in the Babylonian Code of Hammurabi. It was still used as a punishment in some parts of the United States into the nineteenth century. It's hard to imagine any of this history was on Van Gogh's mind when he picked up the razor blade.

How much of his ear did Vincent lose? Accounts differ widely and the extent of his ear loss is either minimised or embellished, depending on what we're supposed to make of his character. Was he crying for help or demonstrating something else? Was Van Gogh's self-mutilation, the removal of small chunk of ear, a cry for help? Or did he cut off the whole ear, a far more dangerous thing to do considering the risks of blood loss or infection? Or maybe he was too mad to be making any statement at all.

It was not just Jenkins' and Van Gogh's ears that have made headlines. In a play on words with the eighteenth-century conflict, in 1992 a political battle became known as the War of Jennifer's Ear. Jennifer was a little girl who waited a year for an NHS operation to cure her glue ear. The Labour party used her story in a political broadcast and it led to a storm over the ethics of using a young girl for political purposes. The Labour Party lost the election. I'm starting to see a pattern here: invoke ears and you might just lose.

In 1997 Mike Tyson bit Evander Holyfield's ear in a boxing match. The whole event was handy deflection for Tyson, given that he was also being accused of controversial behaviour outside of the boxing ring. His ear-biting was the only one of his escapades that was talked about for weeks.

Van Gogh's ear has become at least as famous as any of his art, and is the only thing that many people know about him. Using DNA and replication techniques, a living replica of the artist's ear has now been put on display in an art museum in Germany. His ear became art, just like his sunflowers.

Tsarevich Alexei Nikolaevich's blood

As Leopold, Duke of Albany stepped out onto a tiled floor in 1884, he slipped and fell. He hit the ground and the crash caused the vessels within his knee to break open. Worse, he also hit his head and the blood vessels in his head ruptured too. We've all slipped on a wet floor and had our pride bruised, but most of our blood vessels have a way of stemming the bleeding. But for Leopold this was bad – his blood would not stop flowing.

The Duke of Albany was Queen Victoria's son. When he was born in 1853 the Queen had famously used the exciting new anaesthetic chloroform, the agent developed by James Young Simpson. The use of chloroform at Leopold's birth, whilst an interesting side point, had no effect on the trouble in his veins.

Queen Victoria's royal family had a problem. Their blood, blue as it was, was not clotting properly.

Normally, when the vessel wall is disrupted a chain of events is triggered. A cut of the skin or a disruption within the body signals that a clot is needed to act as a Band-Aid and prevent blood loss. Twelve proteins in the cascade, called factors, set each other off in turn, each catalysing the activation of the next, ultimately producing fibrin. Fibrin plugs the hole, recruits other factors and stops the bleeding. If anything interrupts this chain it is going to affect clotting to one degree or another.

Leopold had a defect in the gene that codes for the Factor IX protein within the clotting cascade. Without this clotting factor, his bleeding was not stopped. The smallest of injuries can lead to blood loss, which over time can be fatal. On this day there was no fibrin clot to shut Leopold's blood vessel's bleed off. Leopold had suffered with this throughout his life but this time, at age thirty, the bleed into his brain was the worst yet. The blood seeped in and pushed on the brain tissue. The Duke died a few hours later of a brain haemorrhage.

When Victoria was born in 1819 there were no obvious signs of the disease within the family. It has long been understood that the family's haemophilia started with Queen Victoria's own spontaneous mutation. Some have even said this is evidence that Victoria was not even legitimate. More recent evidence suggests that the disease could be seen in the ancestors of her mother. Haemophilia was starting to be recognised as an inherited bleeding condition in the nineteenth century and although it is not restricted to royalty, it became known as the Royal Disease.

Haemophilia is not one irregularity but is a group of disorders characterised by clotting problems associated with a missing or broken part of the clotting cascade. Haemophilia A is the result of a break at Factor VIII and haemophilia B a break at Factor IX. These are X-linked recessive genetic disorders where the gene that encodes the proteins are found in the X chromosome. Males have one X chromosome and therefore are more susceptible to the disease. If the gene is present, there isn't another X chromosome to compete with it, as there is in females, and the likelihood is a haemophilia phenotype.

After the first year of life when children start to walk, bump into things and fall over, we start seeing the problems associated with haemophilia. Spontaneous haemarthroses are painful and inflamed and so the range of movement and function are greatly reduced. The disease can be debilitating and so young Leopold grew up with the threat of bleeding to death from the slightest injury hanging over him. In the winter of 1884, the Duke was suffering with painful joints and his doctor suggested that an environment more conducive to healing might be of help. England is cold and wet (he wasn't wrong) so the Duke should get himself to their house in Cannes in France where a spring warmth might do him good. It was there, away from his pregnant wife who he left at home in cold wet England, that he fell. This was a simple accident from which most would recover easily.

Queen Victoria passed on the trait to three of her nine children. As her children went far and wide across Europe, making important strategic marriages, they took haemophilia genes with them. Victoria was known as the grandmother of Europe because her family married into royal houses across the continent. She had a grandson, Friedrich, who also died of blood loss when he was only two years old. Another grandson, another Leopold, died at thirty-two, and Maurice at twenty-three. Leopold's daughter Alice inherited the gene and her son Rupert was born with it.

In 1904 The Tsarevich of Russia, Alexei Nikolaevich, inherited the gene from his mother Alexandra, granddaughter of Queen Victoria. When the boy was born, the umbilical cord bled and bled. It was alarming for his parents to witness. It was clear that the boy had the bleeding disease that ran in his family. As he grew he suffered from bruising, bleeding and joint pain but it was all kept quiet from the public. As ever, keeping secrets led to rumours that filled in the gaps. Alexei was murdered by the Bolsheviks in 1918 when he was just thirteen years old.

The disease did not stop with the teenage Alexei. Elsewhere

in Europe, King Alfonso XIII of Spain married another grand-daughter of Queen Victoria, Victoria Eugenie Battenburg of England. They had boys who carried the gene and it was not until they had a fourth son, Juan, that they had an unaffected heir. The law in Spain said that only a son without physical compromise could become the monarch. Can't think why they came up with that idea . . . we're looking at you, Habsburgs. Public opinion turned on Victoria Eugenie, blaming her for bringing bad blood to the throne, and the monarchy struggled to keep the confidence of the people.

With all families that pass on the genes that lead to haemophilia there can be tragic consequences. With the royal families there could also be a profound political impact. That Alexei's parents spent considerable effort looking after the Tsarevich made them take their eye off the political intrigue that ended all their lives. How would Russia and the boy have fared had he lived and become Tsar? It was a perilous existence with a condition hanging over him that meant a straightforward injury could result in death, just like it did for his uncle, Prince Leopold, Duke of Albany.

Kaiser Wilhelm II's arm

The body parts of the far-flung family of Queen Victoria had a significant impact on the twentieth century. Most significant of all those body parts might be Kaiser Wilhelm's left arm. He had escaped the gene that caused haemophilia but had his own medical condition to face. As Kaiser, Wilhelm was the last German emperor and king of Prussia. His and Germany's military support of Austria-Hungary would lead to the First World War. Wilhelm's troubled upbringing and how he dealt with his weak withered left arm have often been blamed for the carnage of the early twentieth century. Was the Kaiser's left arm another body part that started a war?

Wilhelm was born in Potsdam in Prussia in 1859, the eldest grandson of Queen Victoria. He was second in line to the throne. His mother was Victoria's eldest daughter, Princess Victoria, known as Vicky. She was only a teenager when she married Wilhelm's father and was soon pregnant. The day of the birth was traumatic and difficult. The baby was in a breech position, meaning it was presenting feet first. Vicky struggled and a doctor had to intervene. She was given syrup of ipecac, a potent emetic that would make her vomit. I can't help but think that was concentrating on the wrong end. She was also given a mild dose of the anaesthetic chloroform. Time was running out to save the child and something had to be done. The baby's bum came first and the doctor pulled out his legs. He then forced the baby's left arm out by pulling hard on it, damaging in the process the upper trunk of the brachial plexus, the array of nerves at the top of the arm. As he pulled Wilhelm's arm, the nerves were stretched and pulled apart. The baby was born but made little movement at first. They tried hard to revive him, resorting to heavy slapping of the child until he finally cried. It was the first of many significant and aggressive medical interventions for the child Wilhelm. The birth had left him with a permanently weakened and shortened arm.

This is called Erb's palsy. The condition occurs when the nerves of the arm are damaged during a difficult delivery causing shoulder dystocia. The baby's shoulders become lodged once the head has been delivered and the arms are forced, causing the nerve damage at Erb's point, the place within the brachial plexus where the C5 and C6 nerve roots come together. Wilhelm could hardly move it. Erb's palsy is named after the doctor who first described it, Dr Erb, himself also a Wilhelm. The suprascapular, musculocutaneous and axillary nerves are most commonly affected by an Erb's palsy. These nerves have multiple motor and sensory components, so damage to these nerves means the muscles they should supply are not properly innervated, resulting in the wasting away of the deltoid, biceps and brachialis muscles. Throughout his life, when Wilhelm was photographed, he went to great lengths to hide his weakened arm, concealing it behind his back or in a pocket. He was embarrassed by the arm and not many got to see it wasting underneath his clothes.

Three days after his birth, the baby was not moving his left arm as he should. He was given cold compresses and massages in salt water. As the months went on, his arm did not grow as expected. Vicky wrote to her mother Queen Victoria, describing how they were attempting to correct the problem. One treatment was the slaughtering of hares and laying the fresh, still warm carcass onto the withered arm. This was supposed to transfer the vitality of the warm blood onto Wilhelm's arm and promote growth. After all, nothing says vitality quite like the corpse of a dead hare.

Vicky blamed herself and tried even more disruptive treatments for the boy, though nothing worked. His right arm was tied down so that he was forced to use only the left. This just led to accidents as he had no functioning upper limbs. When Wilhelm put his left hand on a drum, they hoped this was an expression of interest and so they tied the drum to his body and strapped a drumstick to his hand to encourage him to bang his drum. He was put into machines that stretched him and braces designed to prevent the head tilting towards his stronger shoulder. He was held down and given electric shock treatments, daily. Vicky wrote to her mother, telling her that the boy became irritable, volatile and angry. I can't imagine why his emotional growth might have been somewhat stunted.

Wilhelm was taught to swim and ride and shoot like other little princes but it was a difficult schooling. He found it hard, especially on a horse. After much hardship he was able to do these activities in public, and fortunately there was always a footman on hand to cut his food or catch him as he fell. As Wilhelm grew he went to military school, which he enjoyed, soaking up the ideas of the military, the ruling class and nationalism. When Wilhelm did become the Kaiser he was an insecure man who tried to compensate by being outspoken, mostly in the championing of Prussian nationalism. He did, in the end, bang his drum.

In 1888 his father became Kaiser but he died only a few months later from throat cancer. Young Wilhelm became the new ruler. Almost straight away he fell out with everyone around him. At home he was making social and educational reforms but his belief in the might of Prussia was an issue for the leaders of Europe, cousins and distant relations. Completely lacking in tact, he annoyed them all. In England his cousin George V was on the throne. In Russia his cousin Nicholas was Tsar. When his cousins were such strong examples of what a ruler should be, it's no wonder he had a chip on his shoulder. His right shoulder that is, a chip would have fallen off his left.

He rejected his mother's English progressive ideas, preferring nationalism. Wilhelm was not keen on the English at all. The doctor who injured his arm at birth was English. The doctor who misdiagnosed his father's throat cancer as benign was also an Englishman. In his good hand he held a grudge, and it was a big one, overlooking the fact that his injury had come only in the process of saving his life, as well as his mother's. Wilhelm's grudge manifested in how he treated those around him. He was antagonistic towards his cousins around Europe as if he had something to prove.

When the Serbian Black Hand members were found to be behind the murder of Franz Ferdinand, Austria declared war on Serbia, bringing Wilhelm's Germany with them. Russia moved to help the Serbs, and Wilhelm, who went down as chief amongst the warmongers, went to war with them too. Everyone knew the Kaiser as bungling and arrogant. So with his tactlessness having annoyed the British and French to the West and Russia to the East, he and the world faced war.

In November 1918, at the end of the war, he fled to the Netherlands where he remained in exile after abdication. The German monarchy was over and millions were dead. He hoped Hitler would restore his monarchy but he himself died in 1941 of a pulmonary embolism. The young man with an incurable disability grew up with a grudge so big that it shaped the whole world in the first half of the twentieth century.

More recently, research has suggested it might not have been a simple Erb's palsy that embarrassed the Kaiser so much. It was perhaps an intra-uterine growth restriction. He was born small and thin and sometimes a placental insult, which might mimic an Erb's palsy, can occur. More recently doctors have suggested that his traumatic birth might not just have resulted in a damaged arm but also a hypoxia which affected his brain and made him hyperactive and erratic, with limited social skills. This diagnosis is possible, but like the idea that the war started because of Wilhelm's Erb's palsy, perhaps they too were trying too hard to explain the Kaiser leading Germany into war.

According to the writer Christina Croft, the Kaiser making up for his disability by waging war was not really what was going on with Wilhelm at all. According to her, this was all propaganda, spread about Wilhelm by his enemies, to help justify the First World War. He was, to be fair, depicted as a grotesque figure, sinking his teeth into Europe, someone to be loathed and feared. Croft claimed that in fact he was a gentle man, who fought hard to prevent war and was much misunderstood – a far cry from the warmongering madman we believe him to be.

Every figure who has been painted in a bad light will have a rehabilitation book to their name. Perhaps Wilhelm was like us all and somewhere in between a hero and a villain, which is more likely. Whatever his true character, the story we have is that his arrogance and conceit took the world into the bloodiest and darkest of wars, all because he had a point to prove. He wanted to show everyone that his wasted arm did not prevent him being a proper man or leader, yet now he is most often remembered as one of Germany's worst.

Queen Victoria's armpit

The armpit is an anatomical space, a hollow, and it seems strange to call it a body part but it does have a function. Known technically as the axilla, it's a busy place underneath the shoulder joint, bringing together lymph nodes, nerves, blood vessels and muscle attachments. It's a balmy spot, useful for warming up cold hands on a winter's day. Or even cooling the body of someone with heatstroke by placing cool packs there on account of the large vessels that run underneath the skin. It was here in the axilla that Queen Victoria felt a painful lump growing.

Only ten years before, in 1861, French chemist Louis Pasteur published his germ theory. His relatively recent idea that microorganisms caused disease was taking time to catch on.

Before that it was thought that disease was spread by rotten air (miasma). Decomposing matter filled the air with deadly particles, or so the theory went. Ever since the ancient Greeks wrote about it thousands of years before, miasma was identified by foul smells hanging about. The better hospitals aired the wards daily as a response, opening windows and allowing for ventilation, ensuring miasma be flushed out. It was a start, but hospitals were still risky places. Dying after an operation was so common it even had a name: hospitalism.

People understood that disease could be spread between people through contact or proximity, but the ultimate cause of sickness was thought to be visible things like witches, earthquakes, and comets . . . or it was sent by God. Germ theory instead stated that microorganisms – tiny, unseen living cells – were the culprits. In the Queen's warm and sweaty armpit, the bacteria were having a party.

In 1871, the Queen was staying at Balmoral, her castle in Scotland, when the annoying swelling in her armpit became too much to bear. The abscess, which had become the size of an orange, made it difficult for her to move her arm. Victorian surgeon Joseph Lister was called upon to come from Edinburgh. It would have been quite the trip for Lister to make to Balmoral Castle with his carbolic acid spray machine, but he couldn't say no to the Queen.

What was happening in her armpit was an abscess, a

collection of pus. Cutaneous abscesses develop under the skin, commonly at hair follicles, known as faruncles, and at sweat glands. When a collection of boils under the skin join together they make a large painful mass, a carbuncle.

Usually a skin abscess occurs when bacteria, most commonly staphylococcus aureus, gets in through a break in the skin such as a cut or along a hair follicle. Staph aureus is the leading troublemaker but it normally lives quite untroubled on the skin or mucosa. It has a collection of adhesion proteins on its surface that attach to proteins of the extracellular matrix and when it gets into the body where it shouldn't be, it can stick around.

When an infection is first encountered, the immune system will respond by mobilising white blood cells to the area. There they work to break down the bacterial infection, creating the pus. Pus comprises of immune cells, bacteria and dead skin. Pus, technically named liquor puris, builds up in pockets with a wall built around it, pushing out the skin, which can become tight. The blood that has rushed to the area makes it red, swollen, hot and painful. These are the signs of inflammation. If you were to put your hand into the Queen's armpit and palpate the abscess, it would feel firm and round with the pus fluid inside. The Queen would probably have scratched, rubbed and prodded it as it irritated the skin under her arm. Nothing like that really helps an abscess, but rather makes them angrier. Without a knife to cut open the pocket and release the pus, the abscess can grow, becoming more problematic, more painful, as Queen Victoria's did. When movement of her arm became impeded, it was time to call for help. No one was better prepared to help lance the wound and prevent further infection problems than Joseph Lister, who was working on an answer to the problem of suppurative wounds just like this one.

Lister understood that suppurative wounds – that is, wounds that ooze pus – could kill and had been working on a way of preventing infection in his surgical patients. Lister studied the germ research of Pasteur and understood that should a noxious substance come into contact with the minute living organisms it could prevent infection. He developed the nebuliser machine to spray such a noxious substance, calling it the donkey engine. It sprayed the area being operated on with carbolic, a fine misty spray that killed any bacteria it touched. At varying concentrations carbolic goes from bacteriostatic to bacteriocidal. That is, they go from keeping bacteria at bay to killing them completely by destroying the cells. He adapted his machine to spray the perfect concentration. In the donkey engine the carbolic acid, also known as phenol, is deployed using steam created in the base. He sprayed the yellow mist with its distinctive pungent aroma around the room and accidentally into the Queen's face.

With her hand up above her head to provide the biggest field to work on, Lister sliced into the abscess and let out the collection of pus. It must have smelled terrible but he probably wasn't going to tell the Queen that.

It was not the first time that Queen Victoria had turned to the latest medical technologies. When her youngest son Leopold was born she used chloroform to ease the birth. She had a habit of giving the thumbs up to ideas from the new industries of her Empire. Whilst the development of anaesthesia was impressive and useful it did lead to difficulty at first. More patients were surviving the surgery but were dying of suppuration of the wounds. Surgeons had more time to dig about inside the body whilst the patient was under, but the risk of infection was greater. It was not until Lister added antiseptic techniques to the anaesthetic that big changes happened in the field of surgery.

Lister published his findings on the use of carbolic acid to reduce surgical wound infection in the medical journal *The Lancet* in 1867. As so often in the field of medical advancement, Lister encountered opposition to his work. Those against it included James Young Simpson, the doctor who had brought anaesthesia to the field of surgery in the form of chloroform. The Queen may have approved of both of the doctors' work, but they didn't necessarily approve of each other's. Lister did have the upper hand in one regard. After Queen Victoria survived the surgery, Lister would say that he was the only person to stab the Queen and get away with it. Others tried to assassinate Victoria but nobody else got as close to her with a knife as Joseph Lister.

Ignaz Semmelweis's hands

Outside a hospital in Vienna a sign points visitors to the Semmelweis Women's Clinic. It is named after the nineteenth-century Hungarian physician Ignaz Semmelweis. There are statues of him elsewhere, plus a university and even a minor planet that bears his name. His likeness has been seen on stamps and he is remembered as a pioneer. This would all be very surprising to Ignaz Semmelweis, who could not have imagined he would be remembered with such fondness two centuries later. By the time his life-saving ideas about antiseptic behaviour caught on, the doctor had died, beaten and ridiculed in an asylum. All of this because he suggested that doctors should probably wash their hands.

Semmelweis worked at a maternity institute that allowed

underprivileged women to attend for free. They did so with the knowledge that they were used as practice patients, to help the doctors and nurses to train. As he walked the wards, dissecting cadavers one minute and delivering babies the next, he became preoccupied with the study of a worrying problem. Puerperal fever was killing new mothers at an alarming rate.

Puerperal fever developed a few days after childbirth. The mother felt pain and cramping in her abdomen. Her temperature would skyrocket and her breathing and heart rates would also rise as her body tried to fight. There was a chance the infection within the uterus or the surrounding tissue of the reproductive tract could kill the mother, and it did so, often. Puerperal fever was the single most common cause of maternal mortality in the eighteenth and nineteenth centuries. Semmelweis tried to understand the problem he was seeing at the hospital. At first he thought the issue was overcrowding, but the numbers did not add up. The climate and environmental factors did not give any clues either. In 1847 he had a breakthrough. A doctor and colleague called Jacob Kolletschka died after being cut with a scalpel that he was using to dissect a cadaver. Without his knowledge, the cut delivered pathogenic microorganisms into the wound, where they found warmth and food, an ideal environment for reproduction. These little life forms grew in number and got into Kolletschka's blood. His temperature rose, he felt feverish. His joints and head ached.

Semmelweis recognised that the fevers and symptoms his friend faced before the end were remarkably similar to the symptoms that the mothers on his wards had suffered when they died of childbed fever.

When doctors were called to birthing mothers they came from other tasks within the hospital, often the morgue where they dissected the dead. From there they marched through to the maternity wards and stuck their dirty hands in to help. Nowhere along the way would they stop to wash their hands. Why should they? Miasma was the cause of contagion, not what the doctors might be carrying on their hands.

Semmelweis noted that women were more likely to contract childbed fever if they had been attended by a physician or medical student as opposed to midwives. Moments before they helped deliver the babies, doctors were arm deep in corpses. The midwives, he realised, never touched the cadavers – that was the link. The mortality rate rose only where doctors had moved what he called 'cadaverous particles' from the deceased to the birthing women.

He recognised that, for whatever reason, the incidence of childbed fever could be reduced significantly if attendants would simply wash their hands. Such an everyday behaviour for us now, it was not always the case and was even frowned upon and rallied against. One might be forgiven for thinking that you would be in better hands under the medical care of highly educated physicians with cutting-edge medical ideas compared to the midwives. It was not the case though. When it came to postpartum fevers, midwife-attended street births were far safer than the risk of infection from a doctor's dirty hands.

When Semmelweis tried to introduce a system of hand-washing between patient visits, he was ridiculed. The powers-that-be within the hospital even went so far as to ban the behaviour. They refused to believe that they were the cause of the fever, unintentionally or otherwise. How dare he accuse these enlightened men of killing women. Semmelweis would tell anyone who would listen about his hygiene ideas and tried to educate his colleagues, but ideas were not enough. He had no specific scientific theory to back up his ideas. Germ theory was not yet established to explain his cadaverous particles. He continued to try to push his ideas, but nothing would stick – except the germs, that is.

Hand hygiene was laughed at but also seen as a threat. Semmelweis and his family were even accused of being involved with the revolutionary fervour that was sweeping across Europe, discrediting him further. He was portrayed as a troublemaker and his work was rejected. He became nervous and depressed. In open letters he sounded bitter and aggressive. He broke down and started to drink. How else could he cope with the rejection when he knew that his ideas could save many lives? One day, Semmelweis's colleagues took him for a visit to a mental asylum and as he stepped inside they locked the door behind him. Confined to a straitjacket for his last days, he was beaten and died there when he was only forty-seven years old. It is likely that a wound on his hand that he sustained when he was struck, succumbed to infection. It became gangrenous. The skin would at first be pale and then purple before turning black as the tissue died. The infection moved up his hand and arm and into his blood. One of the first men to understand the importance of hand-washing died due to an infection on his hand. The mortality rate amongst the mothers in the hospitals went up again.

It was not too long before Semmelweis's ideas were vindicated. Louis Pasteur published his theoretical works on germs, working in a laboratory, and Joseph Lister focused on the clinical aspects of prevention and treatments, working with patients. It is remarkable that Semmelweis's work was not taken seriously at the time. He was not the first doctor to question the theories of contagion, particularly in the field of obstetrics and puerperal fever. Fifty years before Semmelweis, a doctor in Scotland saw a link between the attendants at childbirth and subsequent cases of childbed deaths but was also shunned for a similar insinuation. Alexander Gordon acknowledged that doctors and midwives carried the infection between patients, including, he admitted, himself. In his *Treatise on the Epidemic Puerperal Fever of Aberdeen*, published in 1795, he even listed the doctors and midwives in attendance when illness and death occurred. He published that thirty-year-old Elspet Fife had been attended by a Mrs Keith on Windmillbrae in the city in March of 1791. Elspet fell ill but was cured. Later John Duncan's wife (her name was not recorded) was also attended by Mrs Keith and fell ill after giving birth. She died on the seventh day. Mrs White of Printfield was recorded dead five days after falling ill. And who had been in attendance? You guessed it – Mrs Keith. Dr Gordon had been excellent at careful note-taking. The midwives weren't quite so impressed. It was Gordon, they said, whose presence in the city brought the epidemic of childbed fever with him, not them. Unsurprisingly, they became hostile towards him.

Gordon, like Semmelweis, was just trying to make a difference to his patients. He used the classic contemporary cures of bleeding and purging to balance the humours (just what every new mother needs). He suggested that the mother's clothes should be fumigated and the sheets be burned for safety. Gordon had been in the Navy and was influenced by other Navy surgeons and their ideas of cleanliness and hygiene, amongst them James Lind, who we met trying to rid the sailors of scurvy. Ultimately both Semmelweis and Gordon faced the same challenge. They knew they were seeing something, but couldn't explain what it was or get others to believe them. Both had their recommendations for cleanliness ignored.

In the United States another doctor was having similar difficulties trying to convince his colleagues that contagion was something other than miasma-driven. Oliver Wendell Holmes wrote a paper calling for cleanliness. In *The Contagiousness of Puerperal Fever* (1843) he gave credit to Gordon's work.

Like Lind before them, Gordon, Holmes and Semmelweis found that their work was not to be acknowledged and put into practice for decades. To get from theory to evidence to execution continues to take far too long for patients.

Anyway, best wash your hands.

Galileo's middle finger

Within its own glass case, on a museum shelf in Italy, sits Galileo's middle finger. It has been positioned in an upright posture, pointing to the ceiling, just as one might imagine a middle finger would be doing, defiantly gesticulating at his enemies who objected to his work.

This is not a religious relic. Religious relics are very common, more so in some religions than others. Some are real and some are claimed as part of a saint to increase their worth and interest. There are secular relics too, usually dried out, mummified and frequently not even identifiable as specific human body parts that are preserved, displayed and revered in the same manner as their religious counterparts. The Victorians were particularly good at the secular relic. The lock of hair in the locket, for instance, a cherished reminder of a departed loved one as well as a memento mori, a reminder of our mortality and a warning that we should live our lives to the full. In fact, Galileo was practically the opposite of a saint, he was a heretic, an enemy of the church. And it's his middle finger that is stuck up at visitors to the Florence History of Science Museum, or more recently known as the Galileo Museum.

The finger was snapped free from the rest of the skeletal hand years after Galileo's death by the antiquarian Anton Francesco Gori. You can still see the biggest of the metacarpals that belongs to the middle finger, and on top, the three phalanges, the proximal, middle and distal, that get smaller as they go. Surrounding the phalanges, dried skin keeps them together.

Galileo Galilei (try reading that without hearing Freddy Mercury's voice in your head) was an Italian astronomer and engineer, mathematician and natural philosopher. With the invention of the telescope using tubes and lenses, he developed theories of the law of falling objects and of parabolic paths of projectiles. But challenging theories of how God's universe worked got him into trouble. A lot of trouble. Up in front of the Roman Catholic inquisition, Galileo was charged with supporting the Copernican heliocentric system that claimed everything in the solar system revolves around the sun and not the earth.

Aristotle wrongly claimed that the speed of the fall of a heavy object is proportional to its weight. Questioning that theory found Galileo some enemies, but he continued to challenge Aristotle's theories. His middle finger was firmly stuck up. Galileo published his *Dialogue Concerning the Two Chief World Systems* in 1632 and ruffled some feathers. It went against the Church and he was warned that he must renounce his own publication. He refused and for the rest of his days was kept on house arrest. When he died, he was not allowed a burial with his family because of his conviction for heresy.

One hundred years after his death, when Galileo was vindicated, the Church removed his book from the blacklist and allowed his body to be exhumed and reburied with respect. His remains were moved from the small chapel of Saints Cosmas and Damian to a mausoleum in the main body of the church of Santa Croce in Florence, Italy. He was given a respectful burial at last, but not before having a few fingers stolen and put on display. When Galileo pointed at the stars or adjusted his telescope, he used these very fingers. To Gori, Galileo's thumb, index finger and middle finger meant so much to the modern scientific age and should be treated like a relic in the same way the Church treats its saints.

There are many different origin stories of why we stick the middle finger up as an insult. In the west, it is thought of as more an American gesture than a European one, but it far outlives the United States. In Latin even the name of the finger suggested an obscenity, *Digitus obscenus* or 'lewd finger'. In England the myths and legends of the Plantagenets rampaging through France with longbows is accompanied by a common story of them sticking their fingers up at their enemies, Longbowmen who were caught risked having their fingers cut off so that they could not shoot any more arrows, and those with the finger still attached to their hands waved them at the French as a sign of defiance. 'Look, we have our fingers, we can still shoot you with our arrows.' But in 423BCE, a very long time before Henry V was willing his men *once more unto the breach*, Aristophanes wrote a play called *The Clouds* in which the obscene gesture was, well, raised.

The middle finger was a gesture used by the Greeks. In fact, all over the world there are stories of middle-finger movement or display, including Arabs pointing the finger down to the ground, and the Russians, who pull it back with their forefinger. (Obviously I mean the forefinger of the other hand, I doubt it's possible with the same hand and now I imagine you are trying it out.)

Galileo's finger could easily be displayed flatly, so it must be deliberately displayed in this manner. Maybe it was pointing to the stars, maybe he's flipping the bird to say, 'Perhaps I was right all along.'

Robert the Bruce's
and Percy Bysshe Shelley's heart

A church tower on the site of Dunfermline Abbey in Scotland bears the inscription 'Robert the Bruce'. This was the burial site of the rulers of old in the northern Kingdom, including Bruce, an iconic King of Scots who led the first war of Scottish independence against England. He was laid to rest here – well, most of him. Bruce's heart is somewhere else and it made quite the journey to get there.

Robert the Bruce died in 1329. For years a story was told that the revered Scottish king had died of leprosy. That's now thought to be unlikely, but like many a myth of monarchical death, it has stuck. When he died, his body was cut open and his guts removed, as was the custom. Guts cause trouble very fast in dead bodies when the unchecked bacteria-filled faeces leak out. A corpse is much easier to embalm if you whip out the guts. You'll need to bury the guts fast, to keep them from stinking the place out, and Bruce's were buried at Cardross where he had died. His heart was taken to be embalmed separately and the rest of his corpse was laid to rest at Dunfermline Abbey.

Bruce requested that his heart be taken on a crusade to the Holy Land, to be prayed over at the Church of the Holy Sepulchre in Jerusalem. The precious relic was put into a metal urn and presented to his friend Sir James Douglas on a necklace. Off Douglas went towards the Holy Land with his friend's heart in a box. In Spain, King Alfonso XI of Castille had been fighting against the Moorish kingdom of Grenada, and Douglas and his companions joined the Christians in their struggle. Whilst besieging the castle at Teba disaster struck the Scottish contingent. They were cut off from the others, surrounded and overwhelmed. Black Douglas, so named because he brought fear to his enemies, particularly the English, was killed there in Spain. The nobleman's last action before drawing his sword to fight the Moors was to throw the urn and heart of his king ahead of him, towards the enemy. 'You go first, Rob,' he shouted, or words to that effect. Neither Douglas nor Bruce's Braveheart made it to the Holy Land. Their remains were recovered and taken back to

Scotland. Robert the Bruce's heart was buried near his friend Douglas at Melrose Abbey, not at Dunfermline Abbey where his body was interred.

Across in Dunfermline the rest of Robert the Bruce was lost under the remains of the Abbey. Most of Robert's tomb was destroyed during the Scottish Reformation. In 1818, workmen who were building a new parish church on the site of the eastern choir of Dunfermline Abbey discovered a tomb near where the high altar would have been. This was something special. Amongst black and white marble fragments there were two large stones, one a headstone and the other a bigger piece around 6 feet (182 cm). Underneath was an oak coffin and within that a skeleton wrapped in two layers of lead, all covered by a shroud of cloth of gold. There was a lead crown sitting on the skull and looking closely at the skeleton it was clear that the breastbone had been cut open to remove the guts.

All in all, this looked suspiciously like the remains of King Robert the Bruce. There was no evidence left of the soft tissues so it was impossible to know if a heart was included when this body was buried. Over in Melrose Abbey, Robert's heart was exhumed during restoration work in 1920 but it was buried again without any marker stone. It was not until 1996 that an urn thought to be carrying the heart of Robert the Bruce was discovered once again. Investigators said that there was no way of knowing for sure that this was indeed the remains of the Bruce's heart, but Donald Dewar, the First Minister of the first government in Scotland for 300 years, said it did not really matter. At a time when Scotland were pushing for independence, 700 years after Bruce was fighting for the same thing, Dewar laid to rest what is likely Bruce's heart under a marker stone. Whether it was Bruce's heart or not, it represented so much.

What the heart represents has changed throughout the world and history. Hearts have weaved their way into the myths of all cultures. For some it was the very centre of the soul, the seat of emotion, all coming from its rhythmic unstopping beat. Aristotle believed the heart was the main event, the organ from which our lives were governed. Hippocrates and Plato weren't so convinced. For them the brain had more of a say. They all agreed though that a significant heat was given off by the beating heart.

The heart shape has been used as decorative art since the ancient societies. Not really anatomically shaped, the symbol of the heart represents the spiritual nature of the organ, of emotion, affection and love, the place where the soul lives. The Teotihuacan culture of ancient Mexico understood how essential the heart was to life. They knew that to remove the heart meant death and ascribed it to the organ's spiritual force that was extinguished if taken out or damaged. To

Christians, the sacred heart of Christ became an icon both in art and prayer.

We have Galen, personal physician to the Roman Emperor Marcus Aurelius, to thank for much of our early medical understanding of the human body, which became the mainstay of medicine for hundreds of years. He described the valves and the ventricles of the heart as well as the arteries and veins associated with it. In the sixteenth century, anatomist Andreas Vesalius observed the heart in even more detail, and Leonardo da Vinci was the first to draw it with real accuracy.

In 1628, William Harvey wrote a book challenging the prevailing concept that blood was created in the liver and flushed through to the veins to be used up. He concluded that the blood was not replenished but continuously pumped around the body by the ever-present beating of the heart. This, he pointed out, was likely the only function of the heart.

William Harvey was one of the lucky people to get a glimpse of a living heart, still beating within a conscious human. A young aristocrat called Hugh Montgomery had survived a nasty accident that punctured his anterior chest wall. An abscess developed from the ensuing infection and the skin and tissue of his chest blackened and died. Montgomery lived but he was left with a hole in his chest where his heart could be seen beating. He wore a metal plate over the area and went on to live a full life, travelling, fighting as a soldier, marrying twice.

It was not ususual to see holes in the skull after an operation or injury, but a hole to the chest was incredible.

Fantastical as it sounds, this is not the only time we will encounter a permanent opening into a cavity in an otherwise well and functioning human. Later we will see a window into the stomach (see page 160), but in the 1600s Montgomery was so remarkable that he was brought to see the King, Charles I, who had heard of this medical miracle. He had a look at the heart beating in Montgomery's chest and he even touched it. You would, wouldn't you? At least being touched by the King meant that Montgomery would be free of scrofula, The King's Evil, which royal contact was thought to prevent. Harvey's work, including his writing on Montgomery's beating heart, changed our understanding of the function of the organ.

Though we understand today the beating of the heart and the movement of blood around the body, we still rely on the heart as a potent symbol to express our desires and emotions. The organ has always represented so much more than merely a pump. When the wealthy and aristocracy died, it was the heart above all other organs that was most commonly cut out and interred in a special place, whilst their bodies may be interred somewhere special to others. When Richard I of England, Richard the Lionheart, died in 1199, his heart was buried at the Abbey at Rouen whilst his body was sent to Fontevraud to rest with his parents. When Edward I's wife Eleanor of Castille died near Lincoln in England, her body was buried in Westminster Abbey but her heart at the Dominican priory at Blackfriars, along the river in London, to be joined one day by the heart of her son Alphonso. Her heart's burial was commemorated by the building of an elaborate memorial including an angelic statue under a stone canopy with carvings and wall paintings. During the dissolution of the monasteries in the sixteenth century, like so many memorials, it was destroyed and lost.

The French Kings had their hearts removed and placed in glass jars, displayed upon cushions. When the Sun King, Louis XIV, died he was seventy-seven years old and the longest reigning of a long list of French monarchs. His weary heart was embalmed and placed in the Église des Jésuites on Rue Saint-Antoine in Paris. During the French Revolution, Louis's heart was stolen and spirited away. A portion of the heart made its way to England in the possession of Lord Harcourt, the Archbishop of York, who was a collector of such curiosities. In 1848 Harcourt held a dinner party and among his guests was a chap called William Buckland. Buckland was a geologist and zoologist who was known not only for a bizarre teaching style and an array of exotic pets, but also an obsession with eating strange foods. He had an ambition to eat a piece of every creature that lived. Moles and bottleflies were horrible, apparently.

The party, sitting at Harcourt's dinner table, passed around the interesting object in their host's collection. As the small piece of the King's heart came into Buckland's hands he just couldn't help himself. He picked it up and popped it in his mouth. 'I have eaten many strange things, but I have never eaten the heart of a king before,' he said. Hopefully he finished chewing first.

Would you keep a loved one's heart if you had the chance? Gothic novelist and author of *Frankenstein*, Mary Shelley kept her husband's heart in her desk drawer – of course. After her death in 1852 this vital organ was discovered in a box by their son. It seems strange that she would have kept it, wrapped in silk and in the pages of his last poem, 'Adonais', considering that they were not the happiest of couples. At least it wasn't eaten. Lord Byron on this occasion appears to have behaved himself when faced with the barbecued heart of his friend, so there is that.

The poet Percy Bysshe Shelley was only twenty-nine years old when he was killed in 1822. He packed a lot into his three decades, but not all of his years were happy ones. Shelley was sailing across the Gulf of Spezia when his party was hit by a storm. His boat *Don Juan* capsized and all of the men were drowned. Ten days later their bodies were found washed up on the beach at Viareggio. The corpses were bloated and decomposing so much that they were unrecognisable. Shelley had to be identified by his clothes and a collection of poems by his friend John Keats, who had died only the year before, which were tucked inside his coat pocket when he drowned.

Shelley was a romantic poet who also covered political and social issues. Proclaiming himself an atheist got him chucked out of Oxford University. He attended very few lectures anyway; instead he built a lab in his room where he experimented and he read.

Shelley eloped with his first wife, Harriet Westbrook, when she was sixteen years old. In 1814 Shelley visited his mentor, William Godwin. He was a journalist, philosopher, novelist and husband of feminist writer Mary Wollstonecraft. There Shelley fell in love with Godwin's daughter, Mary. Shelley was with Mary in St Pancras Old Churchyard, visiting her mother's grave, when they declared their love to each other. Shelley was still married to Harriet and this time Mary was the one who was only sixteen years old. Godwin wasn't having anything to do with Shelley's affair with his daughter. He chucked them both out and they went off to Europe together. He at least left some money for a pregnant Harriet, which was nice of him. His son and heir was born but Shelley was too busy with Mary in Europe to be involved with all that.

Mary was pregnant upon their return, but they had no funds and mounting debts. There was a lot going on for the young man, but nothing he hadn't created for himself, and Shelley's health took a battering. He had panic attacks and would hallucinate, waking his whole household screaming after nightmares. His doctor suggested that the warm air and climate of Italy might be good for his health. Off they went again.

Back in Europe, Mary had a baby who sadly died ten days later. Meanwhile Shelley had an affair with everyone who turned up in a skirt. Back home his first wife, Harriet, drowned herself in the Serpentine. Mary then had a boy who died when he was three years old. Shelley's life was not going well and his mind and body suffered for it. Even before the boat went down and drowned him, Shelley had not been in the best of health.

When the bodies were washed up on the Italian shore, Shelley's friend Edward John Trelawney found their remains and with Leigh Hunt and Lord Byron, they built a funeral pyre. The flames lapped at Shelley's body, scorching the flesh and consuming the organs, but they noticed that his heart was slow to catch fire. It refused to burn like the flesh around it. Hunt pulled the heart out of the fire and preserved it in 'spirits of wine' to keep it safe. His ashes were sent to Rome to be buried at the protestant cemetery.

Percy Shelley's heart refused to burn because its tissues were not the same as those around it. Over the years, as Shelley was losing his heart to girl after girl, he was losing his heart muscle to calcium deposits. Calcium was building up, layer on layer, over the heart muscle and turning into hardened stone. Shelley wrote that Mary was depressed, suicidal and hostile towards him, but really it was he who had a heart of stone.

Calcification occurs as a pathological process. Calcium build-ups are associated with hypertension, smoking and chronic kidney disease. Calcium is supposed to be in the body. It is mostly associated with the bones and teeth but it is crucial in the function of the contracting muscles, the progression of impulses along nerves and the clotting of blood, to name only a few functions. However, sometimes deposits of calcium phosphates or other calcific salts build up at sites that are not normally mineralised. When body parts become inflamed, a common result is for minerals to deposit in the wounded, inflamed areas.

We can tell by the amount of calcium that has built up in the coronary arteries the likelihood of a heart attack. If the calcium deposits are in the valves, it can lead to valvular mal-function, which is the result of metabolic disease and tends to hit people later in life, rather than people of twenty-nine years as Shelley was. When calcium deposits into the myocardium (the muscle of the heart) it is a sign that the heart has been damaged by a tuberculosis infection. We know that Shelley suffered from tuberculosis when he was younger, so this is likely the cause of his calcified heart. A chest X-ray would have shown a bright signal around the heart, but Marie Skłodowska-Curie had not been born yet to invent it. A calcified heart is usually an incidental finding in an X-ray, or at autopsy, or when trying to burn a body.

Back in England, *The Courier* newspaper printed a comment about the poet's death: 'Shelley, the writer of some infidel poetry, has been drowned. Now he knows whether there is a God or no.' He was not quite yet considered the stuff of romantic legend.

The poets and novelists that loved Shelley in his lifetime wrote about their friend and his death. He became the 'sacrificial genius', the forever doomed youthful romantic poet whose heart refused to burn. Eventually Shelley's heart was buried in the family vault with his son, Percy Florence Shelley, who died in 1889.

Dwight D. Eisenhower's heart

Whatever feeling we associate with the heart, the organ's function is as a muscular pump for our blood, or rather two pumps sitting side by side. The muscle of the left side is thicker and stronger than the right side because its job is harder. The left side sends the blood off around the body, whereas the right side sends the blood to the nearby lungs. The muscle of the heart is unique, made up of fibres that are different from the muscles of movement or those found in the bowel wall, that must be able to keep going, day and night, every minute of every day that we are alive. The heart's muscles are supplied by the coronary arteries that come off the root of the aorta, the main vessel that leaves the heart. An electrical system of fibres running the heart's length supplies the regular impulses for contractions that squeeze the heart and push the blood out.

Whilst the heart can withstand some pathologies, even holes and calcification, it is so crucial that to damage it or the vessels around it can be catastrophic. We don't want that. Yet heart disease is still a leading cause of death, and that's because of our lifestyles.

The thirty-fourth president of the United States of America, Dwight D. Eisenhower's heart changed the world. Eisenhower had been Supreme Commander of the Allied Expeditionary Force in Europe, a five-star general before he became president of the United States. He was impressive, but Eisenhower's heart was subject to the same pathological stresses and strains as all of us. One day in 1955 he felt it falter.

On 24 September that year, Eisenhower was staying at his mother-in-law's house in Denver, Colorado. After a classic American lunch of none other than a burger, he hit the golf course. He didn't feel too good though and suffered from pains in his chest. He took some antacids and continued his game, but a few hours later it was clear something more sinister was going on. In fact his symptoms were getting worse, to the point where he needed to seek medical help. The doctors informed the president that he was having a heart attack.

Eisenhower survived this heart attack, but the news had an immediate impact. Wall Street went into a dive. This president was popular and all eyes were on the boss. With press conferences discussing his health, even down to his bowel movements, now was a good time to better understand this medical problem that plagues the people of the western societies.

Stepping into the light was an academic by the name of Ancel Keys, who had a new understanding of heart disease. Keys was a physiologist from the University of Minnesota who had been studying nutrition. His hypothesis ignored Eisenhower's heavy smoking and instead focused on high cholesterol from dietary intake of saturated fat. He said that a diet low in fat, low in cholesterol and high in carbohydrates such as wholegrains, would prevent heart disease.

Across the Atlantic in Britain, another researcher of diet and nutrition, John Yudkin, had other ideas. His hypothesis laid out in his book *Pure, White and Deadly*, was that sugar caused the linked diagnoses of obesity and heart disease. Meanwhile the sugar industry paid academics handsomely for blame to be directed elsewhere. Yudkin was shut down by louder voices with plenty of financial backing. Ancel Keys' hypothesis that saturated fat caused heart disease won. We took saturated fat out of everything. Fat, the nutrient that made humans what we are, made our brains, insulated the nerves and was indeed wrapped around every bodily cell in a lipid membrane, was now the enemy. We are still paying for that decision, which made a few people and companies very rich. The idea that fat is responsible for heart disease lives on today beyond any rational thinking or activism.

What was happening within the arteries around Eisenhower's heart was the hallmark of coronary artery disease, the development of atherosclerotic plaques. This build-up of lipids, cholesterol, inflammatory molecules and fibrous elements narrows the lumen of the arteries and hardens them. The plaques can also burst and clots can break free, moving through the vessels and getting trapped in smaller vessels, restricting blood flow. The heart muscle is dependent on the blood supply, and that supply cannot be interrupted. Atherosclerosis happens silently and develops over years until an event like a heart attack comes along that can cause collapse or sudden death.

The assumption was that high cholesterol, caused by consumption of saturated fat, was the wrongdoer. During an autopsy, if you open up an artery and pull out the fatty plaque it is easy to understand how fat and cholesterol can be seen as the culprits, imagining that they go straight to clog up the arteries. But it is not as simple as that, and because of this assumption, a massive world-wide disaster has been brewing for decades.

The saturated fat that was stripped from foods provided both taste and texture that needed to be replaced. It was substituted with sugar and industrial seed oils with a different make-up of fatty acid lengths. Before this, polyunsaturated fats were not found naturally in any great amount in human diets. The government's advisors, food producers and the newly formed American Heart Association all backed this change. These new foods were cheap as chips to produce and easily made hyper-palatable meals and snacks with long shelf lives. They could even be addictive. It was win–win for everyone but the consumers. While these foods became cheaper and more easily accessible, rates of chronic diseases like obesity, hypertension, gout and diabetes have gone up and up. Institutions including schools, hospitals and prisons were stripped of natural saturated fats. Despite the new craze for low fat everything, the western world has become fatter and sicker, and now over 90 per cent of all Americans show signs of metabolic syndrome.

Since Eisenhower was stopped in his tracks by his myocardial infarction, acute coronary syndromes, if identified in time, can be treated with thrombolysis (the chemical breakdown of thromboses or clots) and angioplasty (delivering protective stents within blood vessels to open them and keep them from collapsing). These procedures have much improved outcomes when heart disease hits. Prevention remains the big issue and it is dogma-lysis that may be harder to achieve than any thrombo-lysis.

It's all too easy to believe the multi-million dollar advertising campaigns created by a billion-dollar industry. The reality is that vegetable oils are a disaster. Who we choose to believe and trust about our diets is an important question to ask as the world faces a crisis of obesity and sickness. By January 1961, Ancel Keys was looking out from the newsstands from the cover of *Time* magazine, celebrated for his work. In more recent years, he has become a scapegoat for the anti-fat movement's disastrous outcome. As ever it was not as simple as one man with one axe to grind. Keys also believed in the health benefits of a Mediterranean diet, a lifestyle that has become popular amongst long-lifespan seekers.

If it is not saturated fat causing heart disease, then what is it? More recent narratives have concentrated on inflammation, the role of ingested sugars, insulin resistance, adipocytes (the fat cells as an endocrine organ) and the role of vegetable oils, sold as being miraculously healthy despite producing inflammatory chemicals when broken down in the cells and damaging mitochondria. There is even research into the bowel's microbiome as a possible cause of obesity. Whatever the underlying molecular mechanism causing cardiovascular disease, which has clearly become worse since saturated fat was swapped for sugar, the academics at the heart of research, including those who are plant-based leaning, agree that saturated fat does not cause heart disease. As we have seen so many times, medical research takes considerable time to make it into guidelines and the public consciousness.

It has been seventy years since Dwight Eisenhower felt the pains of his heart attack, sending our diets down the wrong path. As for getting on the right track, we may be some time.

Reinhold Messner's lungs

Tuberculosis is thought to have killed one in seven people that have ever lived. The mycobacterium tuberculosis is spread between people by coughing and sneezing in close proximity. Tuberculosis-infected lungs can spend years decaying as the immune system works against the invader. The disease slowly eats away at the body and so for centuries was known as consumption. In the lungs, conditions are warm and cosy, the bacteria thrive and multiply, thereby making humans so vulnerable.

Healthy lungs are incredible, and are central to many remarkable stories of human endeavour. They've helped us to survive in the most extreme environment, including in the thinnest of air at the top of the highest mountain on earth.

In the summer of 1999 I was in a restaurant in the high-lands of Scotland with a friend. It was decorated to look old with black-and-white photos of a family of restaurateurs as if this place had been there for centuries. A picture of a climber hanging on a wall reminded my friend, a mountaineer him-self, that the body of George Leigh Mallory had recently been found on the slopes of Everest. I had a vague idea of Mallory but over the following years I became obsessed with the story of Mallory and his climbing companion Andrew Irvine. Many people focus on whether they made it to the very top of the earth before they both disappeared. My question has always been *how*?

It was a new obsession for me. I read a lot and even attended a lecture by a Himalayan climber. As he told the audience of his route, the rocks climbed and the required kit, I was thinking about the effects of high altitude climbing on the human body. How does acclimatisation work? How many calories would they need to sustain the effort? How does the body respond to the extremes of altitude and temperature? How do you poop with a massive down suit on? When I stuck my hand up and asked this question his reply made the audience laugh. 'Doctors', he said, 'are very strange people.'

Earlier that year, high up on the slopes of Mount Everest, the climber Conrad Anker sparked up his radio. 'Might have to stop for a tea and snickers,' he said over the airwaves to his colleagues on the mountain, and mentioned something about hobnail boots. He wasn't after a chocolate bar, or replacement footwear. These were the codewords to be used if they found something interesting in their search for the missing climbers of the 1924 Everest expedition. Ankor had found a body. Judging by the clothing that was clinging to the body and the boot still on its foot, the corpse looked like a climber from the old days.

The body was mummified, preserved in the cold and altitude. The weather-bleached skin of his back, buttocks and legs appeared like marble, like a statue in the galleries of Florence. Natural fibre clothing had eroded over the previous seventy years. The hobnail boots, an old rope around the waist and the buckles of his plus fours that were sitting above the knees – all these had survived, hinting that this was not a recent death. The team believed this to be Sandy Irvine, but as they studied the contents of the pockets and the labels on what remained of the clothes, it soon became clear that these were the remains of the legendary climber George Mallory.

George Leigh Mallory was born on 18 June 1886. He attended Cambridge and became a schoolmaster. He was twenty-eight years old when war broke out in Europe. To begin with he did not join up, being in a protected profession, but eventually his time came and he saw action on the front. Unlike so many young men, Mallory came home from the war but he was restless. Like others who had seen suffering, death and destruction at close quarters, he felt that there was more out there, more to do, more to give, more than simply returning to pre-war mundane life. He and young men like him had something to prove and so did Britain, who wanted to show continued supremacy in endeavour. Having lost the races to the north and the south poles, Britain wanted to win the race to place their flag on the highest point on earth, the summit of Mount Everest.

Mallory was athletic and determined. His passion for climbing took him to the mountains of Wales, Scotland and Europe and then the call came for him to attempt the highest of them all. He visited Everest in 1921 and 1922. Both expeditions made headway but they didn't make it to the top. In 1922 the expedition was all about oxygen and its potential use as an aid to gain higher ground. Back at home the question of whether oxygen should be used raged on. For some climbers using oxygen as an aid was no different to the use of other technology, like tents or boots. Opponents, who often had never left their own garden paths, disliked the idea of oxygen, saying it was cheating. Using it just wasn't cricket. When the group went back in 1924, they were joined by young Andrew Irvine. He had an engineering mind and was a whizz with mechanics. One of his jobs was to look after the oxygen apparatus. They set out from the north, from the earth's highest plateau, towards the highest point on earth. Mallory and Irvine set off to make a summit attempt and the pair were spotted 800 feet (245 metres) from the summit, but they were not seen alive again. One hundred years on, I'm not sure if we will ever really know what happened to the climbers.

This was Mallory's third attempt and he could not fail again. The young inexperienced Irvine was sunburnt and dehydrated but they pushed on. At Camp 5, Mallory left a note to their companion Noel Odell who was a day behind them. 'To here on 90 atmospheres so we'll probably go on 2 cylinders but it's a bloody load for climbing.' He was writing about the use of the oxygen that had got them this far, but he complained about the bulk of the tanks. Noel Odell was the last person to see the men alive, but he said they were 'going strong for the top'. Whether the accident happened as they ascended or as they triumphantly returned, nobody knows.

To count as summiting the mountain, one must come down alive, so even if Mallory and Irvine made it to the top, Hillary and Tenzing would still be classed as the first to climb the mountain thirty years later. I like to think that in 1924 Mallory and Irvine stood on top of the world, even if they needed a helping hand from an extra bottle of oxygen to do it.

It was clear from the body's position that Mallory had slid facedown the mountain. He came to rest with arms outstretched as if trying to break his fall on the rocks and ice. He came to a stop and was still alive at that point. The lower bones on his right leg were snapped. He placed the left intact leg on top of his fractured right leg in one last attempt to protect it. He must have felt the pain of the break. There was nothing he could do and there was no chance of rescue. Mallory was going to die there on the mountain and soon. He was bleeding from the broken bone, he was starting to freeze and his lungs were filling with fluid.

Even with the use of supplementary oxygen, there was no climbing and surviving this mountain yet.

At rest or attention, standing or sitting, reading a good book or climbing a mountain, our lungs constantly move rhythmically, in and out without conscious effort. The muscles around the lungs contract and relax, pulling air in through the trachea above and pushing it out again.

The trachea is a large tube that runs from the back of the mouth and nose down through the chest and branches off into the left and right lungs. The tubes, called bronchioles, become progressively smaller and end at the sacs known as alveolae. It is here that the gas exchanges take place. Oxygen is moved into the blood where the vessels surround the air sacs. Carbon dioxide and water are expelled with the air as it is breathed back out. This happens about fourteen times a minute whilst at rest, when there is no disease or extra requirement for more oxygen or to expel more carbon dioxide.

As he walked in the high mountains, sensors in Mallory's

carotid arteries detected that the levels of oxygen in the blood were getting low. They tell the chest muscles to move faster, to pull in more oxygen and push out carbon dioxide and water. The more breaths you take, the more oxygen is brought in and the faster waste products can be got rid of. It's a good thing to blow off CO_2, but it can be dangerous to lose too much. Like everything, there needs to be a balance. Other sensors, these ones in the brain, detect when too much breathing has made CO_2 levels too low. Low CO_2 levels are perhaps more important than low oxygen when it comes to control of breathing. When CO_2 levels are too low, especially at night, a sleeping high altitude climber will be woken suddenly gasping for air. High altitude climbers don't sleep well and for this reason, climbers make sure to sleep with their oxygen masks on, filling their lungs with richer air.

In the death zone, the area high up on the snowy slopes of the world's tallest mountain, where oxygen pressure is too low to sustain human life, the percentages of gases within each litre of air are actually the same as at sea level. Blaise Pascal demonstrated in 1648 that it is in fact atmospheric pressure that falls with increasing altitude. High up near the summit, the pressures are very low, but there is another catch. The relative reduction in partial pressure of oxygen within the lungs is greater still because of the amount of water vapour that also shares the space. It was thought for many years that it was not even possible to survive here, where not enough oxygen would get through to the blood with each breath. Even now, it is thought that were the highest point on earth any higher, human life could not be sustained.

When Mallory fell and broke his leg, he would already have been asking a great deal of his lungs, climbing through the death zone. Now he had fallen, his fight or flight response would have kicked in. The corticosteroids of adrenaline, noradrenaline (epinephrine and norepinephrine) and cortisol would have rushed out into the blood, raising the heart rate even more, filling the blood with glucose and raising the respiratory rate. The pain in his leg would have accentuated it further. Increased respiration adds to the problem of dehydration. With the cold and the oxygen pressures being so low, he was not going to live much longer. The weeks of acclimatisation that made the effort easier for climbers' bodies was not going to save him now. Without acclimatisation we would not last long in this environment at all (unless you are reading this at Everest base camp, of course). Time at higher altitudes allows the body to cope better with the lower pressures, increasing the number of red blood cells. Mallory's body would have had far more red blood cells than you and me at close to sea level.

With oxygen cylinders and a few more years of high-altitude experimentation, Edmund Hillary and Tenzing Norgay reached the summit of Everest first in 1953. The question then became: could it be achieved without oxygen?

The answer was yes. In 1978, the mountain was climbed without the use of supplemental oxygen for the first time by Reinhold Messner and Peter Habeler. Messner wrote that he stopped every ten or so steps, collapsing into the snow as his lungs needed a rest. When he was ready again, he would stand up and take a few more steps requiring an enormous effort. At night he would be woken often gasping for air. Habeler wrote, 'Up there every step is torture; every movement becomes savagely difficult.' Messner added, 'breathing becomes such a strenuous business that we scarcely have strength left to go on.' When at last he reached the summit of Everest by the oxygen in the air alone, he felt there was nothing else to do but breathe. 'I am nothing more than a single, narrow, gaping lung, floating over the mists and the summits.'

Messner's group motto was 'By fair means' – presumably that was a dig at anyone who did it carrying oxygen tanks. Ten years later in 1988, the first female climber, Lydia Bradey, went to the summit of Everest without oxygen. Her claim was disputed, however, as she climbed alone and could present no evidence she had made it.

The team who found George Mallory's body buried him using rocks and said their goodbyes. They pulled their modern down jackets around them and headed down the mountain away from the death zone, to where the air was thicker. They carried his belongings down to base camp seventy-five years after the disappearance of the climber. The body of Sandy Irvine remains lost on Everest, and with him his Kodak camera that might tell the story of what happened.

Mallory's and Irvine's remains are not the only bodies to lie where they died on the high slopes of Mouth Everest. The harsh environment claims many and bodies often have to be left where they fell. Frozen corpses are sometimes even used as waymarkers for climbers wishing to make their own summit attempts. With better technologies that support our oxygen usage and temperature control, better knowledge of nutrition and weather patterns, and with easier ways to relieve oneself in a down suit, people will always be attracted to the challenge of climbing to the highest point on earth, by fair means or foul.

Alexis St Martin's stomach

When Louis XIV died in 1715, his stomach was apparently twice the size it should have been for a man his age due to his enormous appetite and huge extravagant meals. Having said that, most men of his age were not the king of France. The stomach is a muscular-walled pouch and has the capacity to stretch as needed. With so much rich food sloshing about in there, the Sun King's stomach clearly needed to stretch considerably.

A century later, Alexis St Martin and his stomach hit the headlines. I'll warn you, it's not a good idea to be eating your lunch for this one. Put the sandwich down.

Alexis St Martin (1802–1880) was a labourer born near Michigan. He was illiterate (which is important to the story) and transported furs along the river between trading posts for the American Fur Company. On 6 June 1822 he was waiting in line at a company store. Right next to him a man was showing off his loaded musket. Taking his demonstration a bit too far, the man accidentally shot Alexis, the gun going off so close to him that his shirt caught fire and the shot made a hefty wound. It was miraculous that St Martin survived the blast at all. Luckily, there was a military fort nearby and the surgeon was called at once. When the surgeon arrived he came across quite a scene.

The wound was just under St Martin's left breast. A large portion of flesh had been blown off. Ribs were broken and portions of lung and stomach were sticking out. The surgeon, Dr William Beaumont (1785–1853), saw that whatever Alexis St Martin had eaten for breakfast was now escaping from the stomach wound. William set about saving Alexis's life – that was the tricky part – but even when that was accomplished the injured man could not eat, as the food fell out of his wound. Instead, he was given 'nutritious injections per anum' (i.e. they fed him via his rear end with enemas) until they managed to bind his stomach well enough to keep his food in. As he recovered, Alexis developed a permanent hole which provided an open door right into his stomach.

St Martin was treated by bleeding, because more blood loss would have helped enormously in this traumatic situation, wouldn't it? He managed to survive the interventions of the medical man, but even when he was well he did not manage to get himself clear of Beaumont's grasp. When it became obvious he was not going to return to labouring and was not, due to his illiteracy, likely to get a desk job, Beaumont took his patient into his own house and gave him employment as an odd-job man around his property. Beaumont thought of himself as a nice guy, doing St Martin a favour, but his actions were far from altruistic. He wanted to keep his patient so that he could keep examining Alexis's highly unusual wound.

St Martin survived the gunshot, but was left with a fistula, a hole that led directly from his stomach to the outside world. If it was not properly bandaged, food and drink that he consumed would just leak out. The wound itself never healed but the body eventually created its own valve formation to prevent stomach contents falling free. For Beaumont this injury was a wonderful opportunity to study what was going on in this otherwise hard to reach organ. Gastroscopy was still a few years away. For now, prodding and poking at a hole into the stomach was ideal for a curious surgeon. The dead were not that helpful when it came to studying working physiology. Animals who were experimented on had a habit of dying. There was not yet a good understanding of how digestion works, so Alexis's stomach was a rare chance to observe the process of digestion. The chemist William Prout (1785–1850) discovered that hydrochloric acid in the stomachs of animals was the main digestive agent, but it was thought too strong to be in humans. How could such a powerful acid be inside our delicate human bodies?

Before food reaches the stomach, the process of breaking food down has already begun. In the mouth the food is cut up in the process of mastication (chewing). There it is mixed with saliva, which contains the enzyme amylase to break down the starch into smaller molecules. From there it is moved to the stomach. The stomach is a sack that is contiguous with the gut. It sits at the end of the oesophagus, the tube from the mouth, where the cardiac sphincter controls entry. A sphincter is a circular muscle able to open and close a hole. From there the food empties into the duodenum, the start of the small intestine, via the pyloric sphincter. Though some refer to their whole abdomen as their stomach, when doctors refer to the stomach, it is in reference to the specific organ. Food is held in the stomach, which can stretch and squeeze as it needs to, mechanically breaking down foods. The walls of the stomach are lined with epithelial cells that squeeze acid and digestive enzymes into the mix to digest food into small enough parts that will cross over into the blood stream.

Beaumont prodded and poked Alexis's stomach hole. He even stuck his tongue through it, saying he did not perceive any acidic taste. Beaumont observed and experimented on St Martin from 1822 for over a decade. As Beaumont was stationed in different places, he moved St Martin with him and continued the experiments. Despite the hole, Alexis was able to live normally and perform the duties the doctor asked of him, but he was a trapped guinea pig. The doctor did not attempt to close the hole for St Martin. Instead he used him to observe the stomach processing different foods. Bypassing the mouth, he suspended meat and other substances into the hole, attached to string so he would pull them out and measure their level of digestion. He discovered that, as in animals, acid breaks down food in the stomach. This was a chemical process more than it was a mechanical one. Whiskey though, would just dribble out.

At one point St Martin ran away, making his way back to his hometown, where he got married and had six children. Beaumont was furious. The boy was so ungrateful, he wrote, it's almost as if he didn't *want* to be experimented on for the rest of his life. 'The boy' was only a few years younger than Beaumont, but the balance of power sat with the doctor over the illiterate labourer, who struggled to find work beyond Beaumont's household.

In the 1830s Beaumont published a book on his findings, *Experiments and Observations on the Gastric Juice, and the Physiology of Digestion*. He had demonstrated that digestion was more chemical than mechanical, but some movement was involved too. He detailed 240 of his experiments, including how milk curdles in the stomach, and the taste of stomach contents, having tried some for himself. As you do. He became known as the father of gastric physiology, laying the groundwork for our understanding of gastric juices and digestion. St Martin does not get a title.

Beaumont died in 1853 when he slipped on the ice at the top of some stairs. St Martin was then picked up by a Dr Bunting, who paraded him around, charging people to come and see the hole in his abdomen. Nobody liked Bunting, and St Martin became known as a sorry drunk. You can't blame him for such self-medication. At least the drink didn't just run out of his abdomen since he had figured out how to effectively dress the wound.

Alexis St Martin died in 1880, more than six decades after his unique injury, but even then the army medics were not done with him. The army medical museum wanted to preserve his stomach. His family refused and instead left his corpse out in the sun for it to decompose so the doctors would leave him be once and for all. Even so, the military surgeons sent a bag asking for the stomach to be mailed back to them. The family declined, sending the message: *Don't come for autopsy. STOP. Will be killed. STOP.*

Jack Kerouac's liver

Now we've digested that delightful story we can pick up the sandwich again and move on to another vital organ. After a good meal – or any meal, to be fair – the broken-down food is moved across the gut wall into the blood for carriage around the body. First, the constituents of the food must get past the gatekeeper organ, the liver.

Blood comes away from the intestines newly filled with the products of digestion, passing to the liver via the hepatic portal vein. Proteins and amino acids are made in the liver, as is cholesterol, the precursor to several hormones. Here glucose is converted into glycogen for storage, the processing of haemoglobin takes place and ammonia is converted into the safer product urea. Factors for blood-clotting are made in the liver as are those involved in the immune system's function. The liver clears bilirubin, which comes from the regular breakdown of used-up red cells. One obvious sign of liver disease is jaundice, a yellowing of the skin and eyes when the bilirubin that the liver is struggling to clear builds up. Inside the liver tissue, hundreds of ducts collect the bile that is produced and it is sent to the intestines where it aids the breakdown and movement of fat, or it's sent for storage in the gallbladder to be used later. You might be relieved to hear we have nothing to add about the Sun King's liver. It seems to be one of his few organs that escaped complaint. We can speculate that it would have been as huge and fatty as the foie gras he loved to consume, if the dead king's stomach is anything to go by.

When I was younger I often called to my mother, 'What's for dinner?' and the reply was usually, 'Wait and see.' My mum is a wonderful cook, so what a disappointment to be presented with a plate stacked high with liver and onions. As a child I reluctantly ate the liver in gravy over mashed potatoes whilst holding my nose because my mother told me it was good for us. I understood that it was good for me but not *how* it was good for me, especially if I loathed the taste so much. It was not until years later I learned that my mother didn't like liver either, but only cooked it for its health benefits. Perhaps that's a mum's job, looking out for us. It is the liver's job too.

As body parts go, the liver is a sizable organ with a big job to do. It performs a large number of functions, from the synthesis of enzymes to the breakdown of toxins, the excretion of bile and the production of proteins. This organ is the home of many thousands of chemical functions rendering chemicals safe or making them useful. It is deep red-brown in colour on account of its significant blood supply, and if you were to touch a healthy liver it would feel smooth – but please don't try this at home unless you have a hole in your side like Alexis St Martin.

The liver sits in the top of the right side of the abdomen below the ribs. If you were to palpate the abdomen of someone with a happy, healthy liver, you might have a tricky time finding it. Once the liver becomes diseased and swells in response, it can be palpated in the abdomen with ease. Remarkably, a significant part of the liver can be lost and it will regenerate to about the original size, functioning as before. The liver is not the filter that many think it is, that is the job of the kidneys. In the liver, there are thousands of chemical reactions going on every minute requiring high levels of vitamins and minerals. Whether holding your nose or not, eating liver, compared to muscle meat, has a good amount of bioavailable vitamins and minerals. You don't need to eat a lot of liver to get the benefit, but that's quite tricky if, like me, you can't stand it. As we are talking about human body parts, and I don't want to encourage you, Reader, to eat human liver, let's move quickly on.

Jack Kerouac liked a Margarita. Or a plain tequila; he wasn't too fussy. His liver didn't notice the difference in the cocktail; it just had to deal with the alcohol. Human livers have a lot to deal with, particularly our favourite self-medication, alcohol. You might think the effects of a hangover are harsh, but it would be much worse without the liver breaking down the alcohol. Though alcohol can be seen as an energy substrate, it is poisonous in large quantities and so the liver breaks it down as fast as possible. The liver can be overwhelmed by large quantities over a long period of time, and in 1969, as humans were heading to the moon, Jack Kerouac died of an abdominal haemorrhage brought on by a failing liver.

'I am a Catholic,' he said, 'and I can't commit suicide, but I plan to drink myself to death.' He kept his word. He was only forty-seven years old when a vessel broke down and blood seeped out, but Jack Kerouac had lived an incredibly productive life, with fifteen books of fiction, non-fiction, film scripts and poetry to his name. He even had time for three marriages. Those who knew him towards the end of his life found him repulsive to live with. Alcohol can do that to a person.

Kerouac's most famous book, *On the Road*, was written after his sixty-three days alone on Desolation Peak. From where he'd been stationed as a fire-watcher, watching for fire and smoke, he had to walk for days to reach the nearest road. When he came down from the mountain he wrote the book in only three weeks on one single roll of paper. *On The Road* is a book of poetic prose that spoke particularly to young men who were known as the beat generation. Truman Capote once said of Jack Kerouac, 'The cruellest thing you can do to Kerouac, is re-read him at 38.' Kerouac, later dubbed the father of the hippies, was made famous by the book's success and tortured by the attention. He found it hard to write or think thereafter, but he never found it hard to drink.

There are different alcohols produced by the fermentation of either grains or fruits or sugars. The one that we drink has the chemical name ethanol (CH_3CH_2OH). This psychoactive substance works as a depressant on the central nervous system, where it depresses glutamate, the excitatory neurotransmitter, and increases GABA, the inhibitory neurotransmitter. Thinking, moving, talking, attention, judgement and memory are all affected, and the effect worsens as consumption increases. Alcohol also increases the neurotransmitter dopamine and it turns even the worst coordinated of us into excellent dancers. So despite all the negative effects on the body, alcohol feels great. As with anything that bumps up dopamine, continued exposure will diminish the effect, meaning more is needed to gain the same benefits. The more Jack Kerouac drank, the more he wanted to drink.

If he was sober enough, he would have noticed a yellowy tinge to his skin and the whites of his eyes as the bilirubin failed to be cleared out by the diseased liver. He lost weight and might have felt his abdomen swell with fluid, a condition known as ascites.

Most ethanol is broken down into smaller parts in the liver with the help of an enzyme, alcohol dehydrogenase, and then it's turned into acetate by aldehyde dehydrogenase. Within Kerouac's liver, the acetaldehyde from the alcohol would have caused inflammation because the stresses damage the biochemical pathways within the cells, and would ultimately have affected the activation of his genes. There followed a build-up of fat and fibrosis, inducing an immune response from his body. In short, alcohol wrecked his liver cells and stopped them from functioning. With so many vital functions performed here, to damage the liver is to damage the whole body. The only real answer is abstinence and perhaps, when it has become so diseased it cannot support life, a liver transplant. That was not an option for Jack Kerouac since the first liver transplant that survived in a new body for over a year was only performed in 1967. Really he needed to address the psychological reason for his drinking.

Liver disease and cirrhosis are not always caused by alcohol. Baffled doctors have seen for years the same pathology without the alcohol consumption. Often physicians simply refused to accept that their patients were not drinking and assumed they must be lying. Yet it's not too hard to find the culprit if we look closely enough. As with alcohol, the problem is consumption the liver cannot deal with. Our diets have changed dramatically over the past hundred years, from saturated fats, meat and whole vegetables, to highly processed foods, sugars and oils that are unrecognisable to our livers. The result is an alarming rate of non-alcoholic fatty liver disease (NAFLD), now even seen in our children. It's the same problem causing so much heart disease: it's hard to convince people that the issue is not saturated fats, which humans have consumed since humans became humans, but the low-fat, fake-oil, sugar-laden alternatives that we are told are healthy by people whose only interest is to line their pockets. Multi-million-pound adverts using major sporting events to 'sports-wash' their products are difficult to counteract.

In 2009 the American medical doctor and researcher Dr Robert Lustig gave a lecture at the University of California that would change countless lives. It has been viewed over 24 million times on YouTube. 'Sugar, the Bitter Truth' told us that NAFLD, caused by fatty deposits of sugar, processed food consumption and metabolic syndrome, is fast becoming the number one reason for the need for liver transplant. Not alcohol and not paracetamol overdose. Fatty liver affects one in three in the United States and 6 million children. In fact, it is the number one liver pathology in children. It is no longer excessive alcohol that our livers primarily need to be concerned about, but our diet.

Fanny Burney's breasts

Formally established in the 1750s but likely attracting people for a millenium, London's Borough Market brings thousands of visitors to Southwark every day. It has been, through all those years, a bustling gathering point where people would trade goods and cross the old London Bridge. The market was not the reason for my visit though – I was on the trail of some body parts. After pushing through the crowds, with their wonderful pork pies, chocolates and artisan delights, I crossed the road in the shadow of the Shard building's contrasting modernity. From there, off St Thomas Street, I entered an old brick church and stepped back in time.

This church had been part of the old St Thomas' Hospital. The boarded-up operating theatre in the attic was forgotten after 1862 when the hospital moved. The hidden gem was not seen again until it was rediscovered in 1956. It is now a museum for those who know where to look.

After climbing up a tricky wooden spiral staircase, I entered into an ancient herb garret in the roof and then went up into the old operating theatre itself. There, a semicircle of stalls looks down onto a wooden operating table. There was sawdust below and a feeling of dread above. This atmospheric theatre is a relic of the early nineteenth century, a time of surgery before anaesthetics. The market traders might even have heard the screams.

Behind a set of foreboding double doors was a women's ward. This theatre would have been used for the women who had the misfortune to need hospital treatment. There would also have been surgical demonstrations here, with students and surgeons looking down on people at their wits' end, in such a poor state that the only option was to go under the knife and risk death. This explains the name 'theatre'. It was a show indeed.

In 1811, surgeons were dissecting criminals from the gallows and any other corpses they could get their bloody hands on. If the patients in the hospital did not survive their illness they might have ended up here being dissected. The surgeons practised on the dead and animals (vivisection) to further their skills. Anaesthesia took the form of a touch of wine or opium and a leather strap to bite on. Germs were not yet discovered, and the four bodily humours still ruled medical thinking. The thought of undergoing an early nineteenth-century operation is chilling.

Frances 'Fanny' Burney was a novelist. In the early 1800s when living in France, she felt a pain in her breast. She consulted the medical men, who found a hard lump and suggested surgery. My own family and friends have been affected horribly by breast cancer, as many of us are, and so perhaps it is fitting to use the breasts as an example of how surgery has evolved. Breast operations have saved many lives, but in the early days surgery often led to anguish, pain and heartache. Fanny's choice was between excruciating pain on the surgery table then, or a painful death later, should the disease spread. The risks were high but she chose surgery.

Nine months after the operation, Fanny wrote an account to her sister. She didn't hold back. The pain in her breast had been worsening, she explained, until she could no longer raise her arm. Under the skin, the breast tissue and the fascia (the muscles of the chest) attach to the arms. They contract and relax, flexing, extending and rotating the arms. Coupled with that, vessels coming from the heart and out to the limbs run through the chest along with the lymph system. Nerves, too, wind their way through this area, and of course the bones could well be affected with advanced breast cancer.

Fanny's operation was not performed in an operating theatre

under the gaze of eager medical students, but at her home in Paris. A bed was placed in the middle of the room and covered with old sheets to prevent the dirtying of precious new linen. Body fluids do tend to make a mess.

Fanny screamed below the knife. She wrote, 'When the wound was made and the instrument was withdrawn, the pain seemed undiminished, for the air that suddenly rushed into those delicate parts felt like a mass of minute but sharp and forked poniards, that were tearing the edges of the wound.' She felt the metal against her breastbone, chopping away at her breasts. The thin cloth handkerchief that had been put over her face did nothing to obscure her view as the surgeon lifted the shining metal knife towards her chest. Nociceptors (pain receptors) and sensory nerve endings fire when meeting a noxious stimulus, like chemicals or heat. In this case it was the surgeon's knife that was cutting the skin. The signal is sent via the spinal cord to the brain where it is recognised as a potentially dangerous insult. The brain replies with a fast motor response, usually to move away from the offending stimulus. The pain alerted Fanny to the tissue damage, in case she had not noticed the surgeon and knife before now, but Fanny could not move away. Instead she lay there as the knife went through her skin and tissues to remove the tumour.

*   *   *

Embedded within supportive fibrous connective tissue and fat, there are fifteen to twenty glands in the breasts called lobes. These are made of smaller lobules where milk is produced. They are arranged in clusters like bunches of grapes. (Don't we just love to equate body parts with foodstuffs?) The breast tissue is located in the pectoral region on the anterior chest wall. Both men and women have breast tissue, but the hormonal patterns in puberty grow women's breast tissue for potential breastfeeding from the mammary glands. There are two major parts to the breasts: the larger prominent breast and the less obvious tail that runs into the axilla. The lymph system is important here too, because it is through lymph that breast cancer cells metastasise.

In 1600 BCE breast cancer was described in the papyrus writings of the ancient Egyptians. Galen turned to the four humours to explain it, theorising that a 'coagulum of black bile' within the breast was to blame. Whilst more is known of the disease now, it is still complex. Most breast cancers develop in the milk duct system. Within the nucleus of cells, DNA forms genes, the codes that create protein. When DNA is mutated, proteins may be faulty. When a mutation occurred within an oncogene, cell division can become out of control. When the growth formed a large enough mass, Fanny could feel it and it eventually impinged on her ability to move her arm. The surgeons cut out the lump. It took months to recover, but she survived. There was no infection and the wound healed. There was no breast reconstruction for Fanny though. That would not come until later, although remarkably in the same century.

The first breast reconstruction was performed in 1895 by a German surgeon in Heidelberg called Vincent Czerny. He had removed a tumour from a patient's breast and replaced it with a fatty lump that he cut from her flank and pushed under the skin of her chest. He was trying to 'avoid asymmetry' in his patient, a forty-one-year-old singer. It was remarkable thinking, to use tissue from her own body to make the breasts equal again, but it didn't last. He went on to shove all sorts of stuff in women's breasts, from injections of paraffin to glass balls, wool, ivory and ox cartilage. In no account have I read the names of the women, not even the first one, the 'dramatic singer' that Czerny wrote about. Perhaps he was concerned about confidentiality; perhaps it did not matter to anyone.

Fanny's account of her ordeal is fascinating if not a little hard to read. It is held in the British Library. For so many others, surgery was not available and we will never know the stories of women who suffered through the late stages of breast cancer, which often caused terrible fungating growths to break through the skin, leaking foul-smelling discharge. Whilst this does still happen, particularly where women are not able to access healthcare, and even in a modern-day London, thankfully breast cancer can be recognised earlier and treatments are widely available to stop that spread from happening.

Louis XIV's rear end

Louis XIV was the king of France between 1643 and 1715. He was on the throne for more than seventy years, but it wasn't always comfortable sitting there. The physicians and surgeons who worked for Louis XIV certainly had their work cut out, especially with the fistula affecting his rear end. Fistulae were recorded long before Louis XIV made them popular, but the thing about Louis's bum, and his health in general, is that everything was very well recorded.

His physicians described Louis's rheumatism, vapours, humours, fistula, insomnia, indigestion, reflux, headaches, fevers and melancholy. They described urinary disorders, night sweats and erysipelas (skin infections). He had vertigo and colds and colic, toothache and gout. He was not short of a problem or three. Everything that the King produced, if you get my drift, was inspected and described in detail to help make decisions about his treatments, be they bleeding, enemas, emetics or tinctures. It turns out you can tell a lot from the contents of a bedpan. The volume, colour, smell and even taste of urine can help detect many ailments, from simple dehydration if the urine is darkened, to stones if bloody and red, to porphyria if blue or purple, to diabetes if sweet-smelling or -tasting. Anyone who has eaten their fair share of asparagus will know that works, though I believe there are people who can't detect the smell of asparagus urine and I would like to be one of them.

In 1686, long before anaesthesia or germ theory, Louis complained to his physician about the painful swelling in his perineum. The perineum is the area between the genitals and the anus. The swelling was sore and soon it developed into an abscess, making it quite hard for him to sit down or ride a horse, or even to stand still. The physicians placed poultices and sugar compresses onto the abscess between his legs. They punctured the abscess to allow the pus to drain and injected it with various substances. They applied red-hot irons too. None of this was successful, and unsurprisingly it was all very painful. Louis had enemas and laxatives regularly to balance the humours that must have been off-kilter for the disease to persist. Louis loved a good enema. He was reported to have undergone two thousand of them. If one was speaking to the

King, even for a serious chat, there's a good chance it was while he was having an enema. A pleasant thought. The enemas did nothing to help the abscess.

With all the prodding, injecting and poking, the abscess soon turned into a fistula. A fistula is a channel that develops from one cavity to another, allowing for the movement of pus and bodily fluids. Louis XIV's fistula was a pus-flowing channel that ran from inside the bowel to a hole in the perineum. He basically developed a brand new hole, just next to the original anus. Today perianal abscesses and fistulae are often associated with Crohn's disease, an inflammatory bowel disease, but they can occur on their own, in the absence of Crohn's, as is believed to be the case for Louis.

Eventually, with the physicians making no headway, the surgeons were consulted. Learned physicians looked down their noses at surgeons, who were not considered in the same class. Surgeons did the cutting and got their hands dirty. Operations risked infection. Nobody wants a suppurative wound in the bum area, but needs must, and Louis's needs were growing as fast as his abscess. Kept a secret from court and the public, an operation was planned.

Charles-François Félix was the King's surgeon. He wasn't just a tooth-puller dragged from the grotty Parisian streets though. He was the son of a lord, brother to a bishop and had some standing in society. Still his rank wasn't up there with the physicians yet, but Félix was about to make a name for himself. He had not performed such an operation on an anal fistula before, so he needed time to perfect his art. A fistulotomy lays open and exposes the tract of the fistula, unroofing it. Unroofing is a particularly grim expression that means the pus was drained out, allowing it to heal. Félix practised on the people of Paris. In hospitals and prisons he found men to prod at and cut who had little say in the matter.

Félix put over seventy-five of them under the knife. We don't know their outcomes or how many even survived, but this was all in service to the King's bum, so pull your trousers down and bite on this leather strap, it might hurt a little. Félix developed two tools to work with. The first was a special scalpel with an elongated curvature of the blade, which became known as the King's probe. The second was a rather useful three-pronged retractor, an instrument that holds holes open, freeing the surgeons hands to operate inside.

Once ready to use his new skills and instruments on the King, Félix got to work. The operation took three hours and was a success. Louis was awake throughout but didn't complain much as the surgeon sat between his legs, probe in hand. And so the King felt better – hooray. I'm sure all the men who were practised on were thrilled. Or those that survived at least. The physicians did not have much to say about it all. They mentioned that they bled the King, so that was helpful. Félix was a hero and was given a title and lands. He immediately retired to his new estate and did not operate again. You would, wouldn't you? Quit while you're ahead, I'd say.

The King's rear end did need more prodding later on, as many of these problems often do, but in general Louis was healed. It was fashionable to copy the King's clothes and hair and mannerisms, and now some were seen with bandaged rear ends to show they too had undergone this trendy new operation. Others went as far as to ask surgeons to cut their bums in mock fistula operations, just to suck up in court. A fistula operation became the next big fashion accessory.

There were other perhaps more important repercussions. Recognition from the King meant surgeons were given more credit and more recognition. French surgeons in particular were now highly regarded.

When the musician Handel was on a jolly to France he heard a song that had been written to commemorate the successful but complicated operation. Handel was the official composer to George III. He turned it into 'God Save the King', to be altered to 'God Save the Queen' as appropriate, and it became the national anthem of the United Kingdom and other countries. There are many origin stories for the national anthem, but here at *Vital Organs*, we're on Team Rear End.

Interestingly for Louis XIV it may well have been a *lack* of surgery that led to his death. Years later when the King complained of pain in his leg, his physicians diagnosed sciatica, but Louis had developed gangrene. A cut or bite had let in dangerous bacteria to cause trouble. Gangrenous legs at first turn white, then purple and then black as the tissue dies. The pain was so much that the King begged for an amputation. By the time the surgeons were consulted and they realised his life may have been saved by cutting off the leg, it was too late.

Samuel Pepys' bladder

The diarist Samuel Pepys was only twenty-five years old in 1658 when his bladder stone got so bad he too underwent the surgeon's knife. As it had been for Louis XIV, it was quite the decision, considering the state of surgery in seventeenth-century England.

Pepys lived in London, where he worked his way to a senior position in the King's admiralty. He kept a diary that has given us wonderful details of a most fascinating period in London's history. Surviving since the seventeenth century, Pepys' influential diaries have made a significant contribution to our knowledge and understanding of his London. He saw civil war, a king beheaded, the restoration of the monarchy, bubonic plague, war with the Netherlands and the Great Fire.

Long before Samuel Pepys had his operation, for centuries learned medics and clergymen had been recording the effects of, and attempting to treat, stones. From Ancient Egypt and Mesopotamia to Greece and India, records have told us of the suffering stones can cause. In 1901 archaeologist E. Smith found a stone at a funeral site at El Amral, Egypt, that had formed inside someone's bladder 5,000 years ago.

By the eighteenth century it was understood that lithotomy, the removal of stones from the body, was such a risky business that specialists were needed. Travelling lithotomists made their way across Europe, cutting for stones.

Above the church, in the old St Thomas' operating theatre where we learned of Fanny Burney's breast operation, there are tools and relics from the lithotomy operations performed in the seventeenth century. Visitors can see posters and books that describe the operations that took place in such a theatre as this. Graphic illustrations show how one might remove a stone in the urinary bladder. It might seem redundant to add the word 'urinary' when talking of the bladder, but stones can also form in the gallbladder. Bladder stones were popular in the days of Samuel Pepys.

Bladder stones form from minerals, often calcium, but not exclusively. The most common make-up of bladder stones is uric acid, build-up of which in joints leads to the painful condition of gout. Pepys first felt the stones when they formed in his kidneys. He was a young student at the University of Cambridge when, after a day out walking, he was struck down with colicky abdominal pain and was confined to his bed for days until the stone was passed. Pepys felt the painful urge to urinate more frequently than usual. There might have been blood in his urine, known as haematuria, making it pink. He was in a lot of pain, suprapubic pain felt in his abdomen, and dysuria (pain on peeing) which is worse towards the end of micturition as the last bit of urine is squeezed out. Nowadays these symptoms are considered along with scans, X-rays and urinary blood detection to make a diagnosis. In the past, a definitive diagnosis could sometimes be made by the passage of a Van Buren sound, a metal rod made for dilating the urethra. If the rod comes into contact with a stone in the bladder it produced a 'sound' as the vibration was transmitted along the rod.

The bladder is a muscular sac that works as a reservoir for urine so that it can be purged all at once rather than leaking out as it is produced by the kidneys. The bladder can typically hold 500 ml of urine, but at around 300 ml, the amount in a can of cola, the walls of the bladder stretch and receptors signal to the brain to seek out a toilet. Rupture of the bladder through simple overloading is rare but can happen. It was said that Tycho Brahe, the Danish astronomer who lost his nose in a duel (see page 61), was killed when his bladder burst as he was too polite to leave the dinner table where his king was giving a speech.

So, Pepys went under the knife. The operation was set for the spring when the light and temperature were best. The operation carried such risks that patients were recommended to make their peace with God before undergoing the procedure. For Pepys' sake, it would have to be over as quickly as possible. He was sitting in a chair, in what we now refer to as the lithotomy position – legs stretched wide apart, knees bent and held down by strong assistants. The approach was to be trans-perineal. In between his legs went Hollier, lithotomist to St Thomas' and St Bart's. At first, a long rod entered the bladder up through the urethra, the tube that links the bladder to the outside world via his penis. Hollier moved the rod around until he located the bladder stone by touch. The stone would then be held in place against the wall of the bladder and a cut made in the perineum and up into the bladder. The stone could then be pulled out through the hole between the genitals and the anus. The wound was left open to allow fluids to drain away and it was covered with a dressing to catch them. Without anaesthesia or antiseptics it was another risky operation, but for Pepys it worked wonders. We are glad he survived to leave us his account of life in 1600s London, including great fires, plagues and his kissing the corpse of a long-dead queen.

The stone Hollier extracted from Pepys' bladder was 2.25

inches in diameter. After the surgery, Pepys was given a cold drink of radishes, marshmallow and lemon. That would help. The surgery left a spare hole in his rear end that took a month to heal, after which Pepys was free of the pain that had plagued him. Complications from the operation could include fevers from infection, fistulae from the bladder cavity, impotence, sterility and even death.

In the next century, in 1725, French composer Martin Marais was so moved by his own lithotomy experience that he expressed it the best way he could, through his music. He wrote the dramatic musical piece, 'Le Tableau de l'Opération de la Taille'.

Posters depicting the operation, like those I saw above the church, tend to stick to the male anatomy. Though not unheard of, women suffered less from lithiasis. When they did, the surgeon reached the bladder via dilatation of the urethra. Still an unpleasant thought but slightly less risky perhaps than having to cut into the perineum. Another reason that there are few examples of female patients undergoing the procedure could be the usual cutting out of women from the medical records and writings of the time.

Pepys maintained his interest in all things bladder and bladder stone-related. In 1602 he even attended a dissection and a lecture at the Surgeon's Hall given by a Dr Scarborough. The lecture specifically touched on urinary organs and they carried on talking on the subject over dinner afterwards. You would think he might have had enough of it all.

Over the years, Pepys had one or two further stone attacks. One doctor he consulted couldn't finish caring for him as, like many Londoners, he died of plague in 1665. Pepys did not have another operation, but a post-mortem on his body when he died in 1703 revealed a good many stones in his kidneys, and the old lithotomy wound in his perineum had reopened as he suffered his final illness.

Luckily, the surgeon had vastly improved Pepys' quality of life. The diarist was so grateful for the procedure that he kept his removed stone in a box. On 26 March each year, he raised a glass and toasted the day his bladder stone was cut free.

Marie Skłodowska-Curie's
bone marrow

Unseen, deep down in her long bones, Marie Skłodowska-Curie's marrow cells were paying the ultimate price for her brilliance. Marie's bone marrow was failing, destroyed by radioactivity, the very thing that has saved so many lives since she discovered its potential.

Marie Curie was the first woman to win a Nobel Prize. She was also the first (and only) woman to win the Nobel Prize twice. She remains the only person to have won the prestigious prize in two different disciplines. Her work, which continues to save lives, took hers at the age of sixty-five.

That work involved the study of radioactivity, a term she and her husband Pierre Curie coined together as they worked tirelessly to free radioactive elements from ore. Working long hours in their leaky shack of a lab in Paris, they extracted only 10 grams of pure radium chloride salts from many tonnes of pitchblende. It was a long and difficult task.

Physics and chemistry came together in the study of radioactivity. Marie learned from scientist Henri Becquerel, who had discovered that uranium salts emitted penetrating radiation without an external energy source. That was fascinating to Marie, who saw that it had many potential applications. She took it up as her field of study and discovered two more elements. One she named polonium after the country of her birth, Poland. The other, which was far more radioactive than Becquerel had shown uranium to be, she named radium. In 1903 she wrote her thesis, *Research on Radioactive Substances*. Together, the Curies and Becquerel won the Nobel Prize, but Marie and Pierre were not there to collect the award. There was simply too much work to be done.

In 1906 Pierre Curie was killed in an accident. On the rain-soaked and busy Rue Dauphine he was crushed under the wheels of a horse-drawn carriage. Marie carried on the work alone and she was offered his chair at the university, becoming the first female professor at the Sorbonne.

Radioactivity's usefulness in the field of medicine for both diagnosis and treatment started with Marie Curie's work. She quantified the effects of radiation on living cells, allowing for the development of radioactive treatments. She also demon-strated that X-rays could pass through materials and interact with photographic papers. That meant that they could show us inside our bodies, outlining human tissues, pathologies and foreign bodies. During the Great War, Marie kitted out trucks with X-ray equipment and travelled between field hospitals where she used the devices to show bullets and shrapnel lodged inside the bodies of the injured. The X-rays meant surgeons could now visualise the injuries within their patients. Marie trained other women to use the vans and she installed X-ray rooms in field hospitals. She took along her eldest daughter, Irene, who she trained and who herself won a Nobel Prize in 1935. What a remarkable family.

She worked for many years with radium and other radioac-tive materials with no protection. Though there was a growing understanding that radiation could be used as a treatment for errant cancerous cells, she refused to believe that the work was causing her own ill health. Even Pierre Curie had been feeling unwell before his tragic accident, but how could they know it was radiation poisoning?

There is no way that our bodies can sense ionising radiation. We can't see the electromagnetic waves. There's no smell to the subatomic particles. We can't touch it and we can't feel it as it moves through our cells, detaching electrons from molecules and damaging DNA. Not until it is too late.

Radiation injury to DNA has a profound effect on cells with a high turnover. Within the bone marrow, haematopoietic stem cells whose job it is to produce new blood cells are at risk, as are the messenger proteins that work to tell the cells how to differentiate. Ionising radiation is radiation of a certain wavelength, with high energy that can break chemical bonds, remove electrons from atoms and molecules, and damage DNA within the cells, causing mutations. As Marie steeped herself in her dangerous work, it was seeping into her cells, unnoticed.

After years of exposure, Marie suffered and died from aplastic anaemia. In the bone marrow, where stem cells normally push out red cells for oxygen carriage, white cells for immunity and platelets for clotting, there was a big problem brewing. Bones are a matrix of proteins and minerals and soft connective tissue within cavities. Marie's cavities, where the bone marrow creates cells, had a pancytopenia – across the bone marrow, none of the cells were being produced properly.

She would have been pale with anaemia as her red blood cells dwindled, making her breathless, fatigued and dizzy. She suffered with headaches. Red cells that normally only last for thirty days in the circulation were not being replaced.

As her bone marrow stopped producing white blood cells (neutropenia) she would have found it hard to fight infections. A lack of platelets (thrombocytopenia) meant she was not clotting. The loss of this vital aspect of physiology was worrying. She bruised easily and alarmingly, was covered in small petechial rashes. Her nose and gums bled. She was losing the precious blood that her marrow, now just filled with fat cells, could not easily replace. Her kidneys suffered, struggling to filter out the toxins they normally remove from the body.

She died at the Sancellemoz sanatorium in Passy, in the Alps of the Haute-Savoie, a long way from the grimy, dark Parisian lab where she had spent so many hours looking for that precious element. In 1995 her remains were taken to the Panthéon, the national secular mausoleum in Paris. There they can be visited, along with the remains of her husband. They are both honoured for their achievements under one or two layers of lead that prevent the radiation in their bodies from infecting those paying their respects.

For Marie to achieve what she did was quite remarkable. Opportunities for women were scarce, almost non-existent in Russian-occupied Poland. Her parents were educators and yet she had to leave home to carry on her own education because women were not allowed to further their studies. In Paris her sister studied medicine whilst Marie worked to pay for it, until she was able to go to university herself, to study physics. She could only have dreamed of the opportunities that some, but not all, women have today.

Marie Curie has historically been portrayed as a loner, spending hours in a dismal lab, dressed in dismal clothing.

She was an obsessive, exceptional and rather eccentric character who sacrificed time with her family and ultimately her life for her science. Thankfully, we now tend to concentrate on her work and what she overcame, rather than her drab clothing and eccentricities.

Now there are fellowships in her name for young scientists across Europe. As we struggle to recruit young women into science fields, Marie Curie is a role model. Her story has certainly made me stop and consider how grateful I am for the opportunities I have had, but it also makes me grateful for lead coats.

In the 1980s, her lab at the university had to be decontaminated. Her notebooks, with their yellowing pages of line drawings, columns of numbers and moments of inspiration, are all still radioactive and too dangerous to handle.

Maria Skłodowska-Curie's bone marrow has found its own place in history. Its sacrifice is a symbol of Marie's endeavour as she brought to us an understanding of these radioactive elements, how they work and what they are capable of. Her bone marrow paid the price for the countless lives that have been saved, all over the world, in the years since her own cells were destroyed by what she called her 'beautiful radium'.

Charles Byrne's bones

Charles O'Brien was born in Ireland in 1761. He came to everyone's attention when he just kept growing. He reached 7 feet 7 inches tall and would have kept expanding had he not died of tuberculosis so young. He had a pituitary tumour that caused his relentless growth and made him famous. He travelled to London to make a living showing off his huge frame, where he was spotted by the anatomists. By then he was known as Charles Byrne. They asked that he donate his body for dissection upon his death. He refused. He asked his friends to make sure that his body was disposed of at sea, but still the undertakers let the anatomists steal his body. It can't have been easy to sneak about with the dead weight of a giant. Suddenly his friends had money to spend. After some years in a private collection, Byrne's bones were put in a museum.

I travelled to London in search of the bones of the Irish Giant, Charles Byrne. My plan was to visit the Hunterian Museum to see his remains amongst a range of human anatomy specimens. Of course, by viewing his skeleton I too was going against his wishes not to be dissected and exhibited. Even after all these years, my eagerness to see this man's bones still felt a little awkward. Luckily for my conscience, the museum was closed for refurbishment. As it happens, the museum have decided, after lobbying from a concerned group including the late Hilary Mantel, who fictionalised Charles in her 1998 book *The Giant, O'Brien,* that the bones of the Irish Giant will no longer be on display in the museum. At last the man's wishes are being respected, over two hundred years since he expressed them to his friends.

In the eighteenth century, surgeons stole dead bodies and asked people with unusual or interesting anatomy to leave their bodies to science. The very thought of going to the afterlife without an intact body was horrifying to many, so much so that dissection after execution was used as a further punishment for severe crimes as part of the Bloody Code. What made Charles Byrne so interesting to the surgeons was his huge size, and it was all due to a tiny clump of cells within his brain.

It might be the size of a pea, but the pituitary gland plays a big part. It controls the hormone outputs of most of the secreting glands. Sitting in the sella turcica, a little hollow in the skull's base, behind the bridge of the nose, the pituitary sends out hormones through the blood stream to distant body parts. The mammary, adrenal and thyroid glands, plus the ovaries and testes, are all told to release their own hormones by the pituitary. The growth hormone somatotropin from the pituitary gland regulates liver, bone, muscle and fat tissues to affect growth, metabolism and body composition. Too much growth hormone coming from Charles Byrne's pituitary gland meant that the boy just kept on growing.

In 2006, researchers took a sample from the bones and did a DNA analysis. It turned out his was a very rare condition. Fewer than 5 per cent of pituitary tumours are caused by a genetic mutation like Byrne's. The genetic anomaly was also found in four families from the same region in Ireland. The researchers found that Byrne and these families in Ireland shared a common ancestor. Now the families can be treated to arrest continued growth, which prevents the painful and even life-threatening complications that can come with gigantism. For Charles Byrne there was no treatment.

There are a lot of other bones and remains besides Byrne's held by the museum, looked after by the Royal College of Surgeons. Of course the majority of those specimens are unlikely to have come with explicit consent to display them either, perhaps having been snatched from their graves. The bones of the giant, however, explicitly came without consent, but his wishes were ignored.

Bones do not decay the way soft tissue does, meaning that we can study, handle and gain new understandings from them, years after the connective tissues, organs and other matter have long gone. Preserved by the museum, Byrne's bones will last far longer than they may have had he been buried at sea or even in the ground.

Bones are made up mostly of collagen. Rather than the solid structure that we imagine bones to be, it is rather a strong porous matrix. Bones are subject to the same insults from chemicals and biological, microorganism processes that cause decay to soft tissues. Bones can resist because of the durability and stability of the collagen associated with calcium and other minerals, the same reason they provide strength to the living body. There are only certain enzymes that will break bones down. If the body is exposed to water or even open air, the porous network inside the mineral outer layer will be invaded by bacteria, fungi and insects, and the bones will begin to decay. Dry and arid conditions, however, mean fewer microorganisms and less breakdown. Without bones' slower rate of decay, we would understand far less about our ancestors.

At Waterloo, very few skeletons from the many soldiers

killed in the battle have ever been found. Was the ground particularly good for the breakdown of bone tissue? Or, like the battlefields of Leipzig, Austerlitz, and others from the Napoleonic wars where thousands of men fell, were the bones stolen when mass graves were raided? For a long time, we believed that the Waterloo bones were taken and used to grow cabbages. Yes, you read that correctly. Bones, ground down into meal, make good fertiliser. When the bones were collected from the battlefields they were shipped to England via the port of Hull. From there they were transported to factories in Doncaster in the north where they were made into fertiliser and sold to farmers across the land. The bones' minerals fertilised cabbages (and other vegetables) to be eaten by those left behind in England.

Amazingly, cabbages were not the only commodity made with the bones of the war dead. Dotted around Europe was a collection of sucreries, French sugar beet factories. During the war, Britain had a monopoly on sugar cane and blockaded France's trade routes with the Caribbean. Sugar was in short supply, so when Napoleon was handed a block of sugar that was made, not from cane, but from home-grown beet, he was taken with it. He gave land to grow the plant and helped establish factories. French sugar beet factories were booming. Chimney stacks at the Mont-Saint-Jean factory threw out smoke from the process of making sugar day and night. It is likely that some of the bones from the Battle of Waterloo ended up here, used in the process of making sugar.

In the village of Mont-Saint-Jean, a bone's throw from the battlefield, the sucrerie is now a hotel offering a bed for the night and a good sugar-filled breakfast in the old factory's central building. I spent a night there, trying to sleep, imagining the rattling bones of dead soldiers being brought into the courtyard, loaded onto carts, to make sugar. Don't let me put you off, but I couldn't even look at the sugar bowl during breakfast.

One Waterloo soldier did manage to escape the cabbage fields of England and the sugar factories of France. His bones remained under the ground until 2012, when he was found during construction of a car park near the battlefield. The musket ball that killed him was still there too, lodged between his ribs. He had a distinctive spinal curvature and carried coins, a spoon and a piece of wood that was engraved with the initials CB. At first there was no way of identifying the man on that information alone. On closer inspection, an F was also visible to make the initials FCB. Historian Gareth Glover cross-referenced the records of soldiers. None specifically fit with FCB, but one Hanoverian soldier had the initials FB, a twenty-three-year-old man called Friedrich Brandt of George III's King's German Legion. In 2015 a museum was built to commemorate 200 years since the battle. The bones we believe belonged to Brandt were put on display, along with the musket ball that had killed him 200 years before. They represent all those who were lost in the battle and those who were lost to the industrial processes used for feeding people after the war.

Remarkably, in the same year Brandt was found, the bones of King Richard III of England, also with a spinal curvature, were found under a different car park. (There's a car park next to the old ruined castle in my village. I might go digging there after dark to see who I can find.)

As on the nineteenth-century battlefields of Europe, archaeologists and historians have tried to find the remains of those killed in previous battles, particularly from the Wars of the Roses. During these bloody battles in the fifteenth century, the Lancastrian army of King Henry VI faced the Yorkist army of the soon to be King Edward IV. Thousands were killed in these battles, but their bones have not been easy to track down. In 1461, in Towton near York, in a battle described as the bloodiest in English history, the Lancastrians were beaten and fled, pursued by the Yorkists. The fate of those who were caught was now laid out for us to see in their bones.

In 1996, construction workers were digging to make foundations for new buildings. They found a mass grave holding the remains of over thirty individuals. These were men who were killed after the Battle of Towton. Over six hundred years had passed between their last stand and the discovery of their remains. In that time, X-rays, radio carbon-14 dating and a greater understanding of the human body overall had been developed. The bones could now tell us exactly how the men met their deaths. The bodies demonstrated some brutal injuries. Great cuts through the bones of the face, poleaxe punctures to the heads, and caved-in skulls suggested these men were executed. They were not blown apart by explosives and gunpowder; the injuries were inflicted by handheld weapons at close quarters. That is what is so alarming about the patterns of injuries, they were inflicted at striking range where the assailants could see the whites of each other's eyes. The remains also revealed how young and strong the men had been. They must have started training for battle when they were mere boys. Nasty injuries had been cared for and healed over the years before their final fight.

Few full skeletons have been found on the battlefield itself because in 1483 King Richard III ordered and paid for the removal of remains from the mass graves to consecrated grounds in nearby churches. Luckily for historians, they did not exactly do a thorough job and researchers have found partial remains.

According to the archaeologist Tim Sutherland, feet and hands were sometimes left where they first were buried, whilst the longer bones and skulls were removed. To Sutherland that suggests that when King Richard III's diggers went back to the battlefield of Towton in 1483, whilst bigger bones came away easily, some of the smaller connective tissues would still have been there, keeping bits of bone together. Collecting these bones would not have been a pleasant job. The bones that did find their way to churchyard graves were at least given some dignity. As bones are slow to decay, more will be found and we'll need to decide how to treat what were once living, breathing human beings with respect.

William Burke's skin

'Dr Knox, FRSE will commence his annual course of lectures on the anatomy and physiology of the human body on Tuesday 4th November at 11 am ... arrangements have been made for an ample supply of anatomical subjects.' The posters were pasted up around the university in Edinburgh in 1828. The former military surgeon Dr Knox's lectures on anatomy were so successful that he had to give two a day. He always had a good supply of dissection material, which helped, but where were all the corpses coming from?

There was a meagre supply from the gallows as surgeons were allowed to claim convicts who had been hanged, for their dissection tables and boiling pots. That was not enough though for Dr Knox. He had to take what he could get from other suppliers. Resurrection men went out under cover of darkness to dig up the newly deceased. Knox paid well for a good fresh corpse. He paid two men named Burke and Hare especially well for the very fresh corpses they brought to his department, without asking any questions about the provenance of the supply. Whenever anyone mentions body snatchers, William Burke and William Hare spring to mind. In fact Burke and Hare were not body snatchers, they cut out the middleman – if we can call a scythe-wielding skeleton in a black robe the middleman. Burke and Hare were murderers.

They did not start out just murdering people. They met as navvies on the canal, became drinking buddies, and Burke stayed at the boarding house belonging to Hare's wife in Edinburgh. An old soldier had been lodging at Hare's house too. He fell ill. So ill in fact that he died. Hare was left with a dead body to deal with and the man's debt, four pounds of lodging payments the old man had left unpaid. So Hare went to his friend William Burke to ask him what to do. Burke knew that body snatchers made money from the anatomists. The answer was not to let the lawmen deal with it, but the medical men, as that way money lay.

The university gratefully received the body for a finder's fee and no questions were asked. Burke and Hare were even told that should any more people die unexpectedly hereabouts, the medical school would be happy to help relieve them of the dead weight. Burke and Hare went for a drink with their pockets a little heavier and came up with a plan. People don't often die unexpectedly, so when an old man in Hare's house didn't just drop down dead as they wanted, Burke and Hare helped him on his way. They earned ten pounds. The pair went on to murder sixteen people. They chose the vulnerable and disadvantaged, people of the streets who might not be missed. They plied them with drink and then one of them pinned the victim down, while the other covered the mouth and pinched the nose with their large hands. The process became known as Burke-ing. The elderly, poor and infirm were easy targets. Few would question their whereabouts when they went missing.

The state also provided medical men with the bodies of criminals. As you could be hanged for a long list of crimes, a further, more shocking punishment was brought down upon the murderers. The threat of dissection was harsh because nobody wanted to meet their maker without an intact body, come the actual day of resurrection. Being cut up was the stuff of nightmares. The public feared their corpses being dug up so much that they came up with different ways to deter snatchers. Manned lookout towers were built to keep an eye over grave-yards, mort-houses were used to lock up dead bodies until they were properly putrefied and of no use to the anatomists. Heavy iron cages were built to sit upon the graves, slowing down and deterring any body snatchers. It was becoming harder for the body snatchers, so Burke and Hare's solution was a useful one for both the anatomists and their pockets.

We might as well be hanged for a sheep as for a lamb, they thought, and the body count rose. Burke and Hare were making a killing. They even killed a family member. Helen McDougal, Burke's wife, had a cousin called Anne whose corpse they sold to the anatomists. There was a young lad known locally as Daft Jamie, a street performer known to students on account of his distinctive club foot. Burke and Hare gave him drink and smothered him. Knox recognised Jamie's foot when his body arrived at the university, so quickly dissected the lad's head and feet before anyone else caught on. Burke and Hare were taking too many risks. At last they were caught when a recently Burked body was found in their lodging house.

Burke was considered the intelligent one, the leader. Hare was given immunity from prosecution for turning on Burke. He testified against his old friend and Burke was convicted of murder. Ironically, he was sentenced to execution and public dissection; specifically the preservation of his skeleton was ordered, to be kept by the medical school.

On 28 January, 25,000 people gathered to watch William Burke swing. It was a short but severe struggle on the end of the rope before he was taken down and placed on a dissection table. Tickets to the dissection had sold out and a riot started when many more turned up than could fit in the dissecting room.

As Alexander Munro, tertius of Edinburgh Medical School, cut up the murdering criminal, he dipped a pen in the blood and wrote a note with it. 'This is written with the blood of William Burke.' Dr Robert Knox was not charged. Hare's decision to testify surely helped Knox, as further investigation would have pointed a finger directly at him.

Nearly two hundred years later, Burke's preserved skeleton, as prescribed by the sentence, is still on display at the University of Edinburgh medical school. It is his preserved skin that surprises visitors to the surgeon's hall museum the most. A book of his murderous exploits was bound in his skin. Without preservation of some kind, the skin would soon start to smell, turn green and slide off the pages underneath. In order to last, and for us to see it today, the skin needed to be tanned.

The tanning process permanently alters the proteins, the structures that make up the skin, so that it becomes hard-wearing and will not decompose so fast. The process starts by curing in salt or salt solution. This is followed by liming or use of a sharpening agent, taking out any of Burke's hairs in the skin sample. Different salts and acids are used to alter the pH. Then the skin is pickled then de-pickled to produce useful leather that will last. Then the leather can be rolled or waxed. I can't help but wonder if the workers at the tannery knew they were dealing with human skin and whose skin it was. Burke's skin leather was then stretched and stitched over the pages to bind the book. This was a deliberate punishment that took time and effort, not a whim like the surgeon writing in Burke's blood.

Burke's was not the only criminal's skin to be used to bind of a book. In fact the practice of binding books in human skin even has a name, anthropodermic bibliopegy. It is not easy to identify a human-bound book by touch, sight or even sniffing it. The Anthropodermic Book Project, a team from the Mutter Museum of the College of Physicians in Philadelphia, has been examining books in libraries and museums that claim to be bound in human skin. They have determined that of thirty-one books so far examined, only eighteen were found to be actually bound in human leather. Identifying human leather is typically a subjective process that comes down to identifying the pattern of hair follicles, which can be distinguished between different species. Whilst DNA evidence could be used, the tanning process likely destroys any useful DNA. Rather than use DNA, the Philadelphia team looked for other surviving proteins under the microscope that can more objectively identify human skin that binds a book, a process known as peptide mass fingerprinting.

John Horwood was a less notorious killer than William Burke, but his body suffered the same fate in 1821. John Horwood was a coal miner's son who was convicted of murder and executed in Bristol, England. A woman ended their short relationship but he kept harassing her . . . because that always works. One day when he saw her with another lad, Horwood threw a stone at Eliza that hit her on the temple. She was OK at first but it was later realised she had a depressed fracture. She was admitted to hospital but developed a fever. The wound was infected and the surgeons had to perform trepanning, where a hole is drilled in the skull to allow the release of pressure and pus. Later analysis suggests the trepanning itself may have killed Eliza, but Horwood threw the first stone. He was hanged, dissected and displayed for over two hundred years as a skeleton alongside an account of his crime bound with his skin. After a stint at the home of the surgeon, Horwood's skeleton went to Bristol University, where it hangs in a cupboard with a noose around its neck, the sign that this was a convicted criminal and his remains are kept as part of his punishment.

Perhaps the most famous example of a book bound in human skin is Holbein's *Dance of Death*, the *Danse Macabre*, which given the subject matter feels appropriate. During the French Revolution, rumours spread of a human skin tannery set up on the outskirts of Paris to handle the excessive number of bodies. If that was really the case, there would be a lot more books bound in the skin of beheaded French aristocrats.

Richard and Ronald Herrick's kidneys

I considered opening this chapter with a story of a steak and kidney pie, but having already grossed everyone out with talk of munching on liver, we will just get straight to it. The kidneys are a pair of organs that filter the blood. At least, that's one way of putting it, but it's a bit more complicated than that. They do not act like simple sieves, letting smaller substances through and keeping bigger substances in. They are more nuanced and, dare we say, more interesting than that. They also control fluid levels within the body.

When blood flows into the kidneys, it enters the glomerulus, the kidney's functioning unit where molecules are filtered out of the blood through a membrane. They move across the membrane according to the different concentrations of those molecules on either side. Higher concentrations on one side will move across to lower ones on the other. Salt levels are carefully controlled and even salts removed early in the process can be reintroduced back to the blood if needed. Waste products are sent off to form urine which is delivered to the bladder via tubes called the ureters. Hormones control the movement of those salts and vary the amount of water that is kept or excreted, depending on the body's needs moment to moment.

The kidneys sit in the back of the abdomen behind the peritoneum, and as such they are vulnerable to trauma. Luckily, if a kidney is damaged, on a rugby pitch say, or during kick-boxing, the other kidney can take on the workload.

The Greeks and Romans, whilst they had knowledge of many conditions, could not get to grips with the slippery kidneys. The swelling or oedema often associated with renal disease was thought to be due to the liver, when it was more likely down to non-functioning kidneys. Oedema, known as dropsy for a long time, was thought of as a disease in its own right rather than a symptom of kidney failure. It was not until 1827 that the physician Richard Bright described acute and chronic nephritis, diseases of the renal parenchyma, and finally correctly identified problems in the kidney as the root cause.

Three hundred years before Bright described renal disease, a sixteenth-century priest was suffering. Martin Luther was a prominent figure in the Protestant reformation. He argued against the indulgences of the Roman Catholic Church and proposed in his Ninety-five Theses that something should be done about it. Along with all of his philosophical writings, he frequently wrote about the discomfort of his troublesome pair of kidneys. He had difficulty producing urine from his stone-damaged organs. Unfiltered substances built up in his blood, causing swellings or even worse problems like electrolyte imbalances that lead to unwanted muscular contraction. Most worryingly, these can even occur within the heart muscle. If the heart muscle starts to twitch because the blood level of potassium is too high, we have problems.

If disease was due to the wrath of God, then God had a bone to pick with Luther. He suffered from, amongst other things, Ménière's disease, vertigo and cataracts. He had angina, arthritis, ear infections and stones in both his bladder and his kidneys, which caused him excruciating pain. The agony, he said, reminded him of death. The Christian in him believed this was sent by God to remind him of life's struggles. They must have been hard to forget.

In 1537, whilst attending a convention of the Schmalkaldic League, he was struck down by one of the worst attacks of renal stone pain. Afterwards he complained to his wife that he had hardly been healthy at all whilst he was away. 'I was dead,' he told her. Many people prayed to God and that meant that he recovered. Or, more likely, the stone passed. The stones damaged his kidneys and he could have done with a new one via transplant, but that was not going to be possible for another 400 years or so. He wrote about his kidney pain about as much as his Ninety-five Theses. The never-ending agony left him quite the antagonistic grump.

Nowadays there are two options if the kidneys are failing like Martin Luther's. The first is dialysis, the removal of substances by artificial means. Dialysis machines take blood from the body, push it though filter membranes and return it to the veins. Blood tests calculate the urea and electrolyte levels before and after the procedure. We can see the levels change as the colour-coded blood results jump from red to green by what feels like magic for someone with failing kidneys.

The other option is to transplant a healthy kidney from a donor. Kidney transplantation changes lives but requires anti-rejection medications for the host body to accept the organ. A functioning kidney reduces the need for regular trips to sit in a hospital by a dialysis machine for hours on end.

In 1954, a twenty-three-year-old American named Richard J. Herrick noticed a severe swelling around his eyes and he felt unwell. Herrick's blood had far more than the expected amount of urea, the waste product excreted by the body in urine. He was suffering from the effects of having too much urea inside him. His kidneys were failing and he was told he had perhaps two years left to live. He got so sick that he could hardly walk. His brother Ronald spoke to Richard's physician – 'Doctor,' he said, 'I would give him one of my own kidneys, if it would help.'

The doctor knew that transplantation experiments had not been going well. Recipients' bodies were rejecting the donors' organs for unknown reasons. The hosts' immune system were recognising the new organs as foreign, sending immune cells to mount an attack and protect the body. There was an inkling that donations from relatives would last longer than organs from strangers. But here was the best part – Richard and Ronald were twins. Identical twins. This might just work.

There was a lot to figure out, chiefly about the physical practicalities of wiring up a new kidney inside a different body, considering vasculature, nerves and lymphatics. Surgeon Joseph Murray performed the first successful organ transplant operation two days before Christmas in December 1954. Ronald did not just give his twin brother Richard a healthy kidney for Christmas, he gave him life. Richard and Ronald both lived, each with one functioning kidney.

Joseph Murray was awarded the Nobel Prize for Physiology or Medicine in 1990. Being able to transplant kidneys has saved countless lives. It is only right that the Herrick brothers are remembered too. So often it is the medical men who become famous and collect awards, but without the patient's brave sacrifice none of it would be possible.

Kidney transplant patients go on to live full lives. In 2019 in the United States the operation was performed a record 24,273 times. You just never know, the person sitting next to you in the restaurant might just have someone else's kidney inside them, filtering their blood and keeping waste products from building up. Without that kidney, the person might not even be alive now. They probably won't be eating steak and kidney pie.

Alfred the Great's guts

In 1620 the medical student John Moir recorded what his professor said about the bowel. 'The intestines', the professor said, 'are comparable to a jester, who unless gravely insulted, remain equitable.' Despite having fought many a battle, Alfred the Great did not die on the battlefield. It was his guts that killed him. Alfred the Great's intestines gravely insulted him into his grave.

Alfred the Great was king of West Saxons and the Anglo-Saxons. He died in 899 having suffered for years with abdominal cramps and bouts of diarrhoea. Medics of recent years have speculated that perhaps the great king had a disease well known to many today, the inflammatory bowel disease Crohn's. The lesions of Crohn's are caused by autoimmunity. For an unknown reason, components of the body's own immune system falsely recognise something on the gut wall as foreign and get to work attacking the invader. The effects can come and go, allowing some pain- and symptom-free periods, but at other times they were debilitating for Alfred.

Alfred is remembered as one of the greatest kings England has ever had. He earned the name the Great through his vision and the masterful qualities as a statesman. He fought and won and brought forward great reforms. Holding his abdomen in pain and having to run frequently to relieve himself, Alfred can't have felt that great. Crohn's can lead to malnutrition and fatigue, loss of appetite and weight loss, and may just have been responsible for the King's death. He died when he was fifty or fifty-one years old.

The intestines are a very long slippery tube that runs from the stomach down to the anus. Foods are digested here and absorbed through the gut walls into the blood. Instincts originate here in the form of 'gut feelings', and that might be more physiological than it sounds. The gut is now known to be able to speak to the brain via a connection between the vagus nerve and the microbiota in the gastrointestinal tract. Down at the microscopic level, micro-projections stick out into the lumen of the gut tubes, increasing the surface area massively, providing more space on the wall for the nutrients to get through and a place to stay for many millions of bacteria who make this warm, mucusy, cosy place their home. There's a lot of tissue in there that can cause trouble when it is gravely insulted.

In Alfred the Great's bowel, patches became inflamed, which were painful, bled and sloughed off mucus. That all has to go somewhere and the only route it can find is to run out the rear end. Alfred's Welsh bishop, Asser, kept a contemporary record of the King's symptoms and modern medics have suggested Crohn's or haemorrhoids. It is of course possible to be afflicted by both, and Crohn's can also lead to abscesses and fistulae like our friend Louis XIV suffered from. Not everyone's medical history is recorded in every tiny detail like Alfred and Louis's. King Henry V of England may well have had a bowel condition but not much was said about it so the warrior king would not be associated with bum disease. How times changed.

Alfred's grandson Eadred appears to have suffered from a similar illness. He endured gastrointestinal problems for years, sometimes confined to his room for days or weeks with cramping abdominal pains and diarrhoea. He could have done with indoor plumbing.

It is thought that Crohn's disease could be caused by an infection of Mycobacterium avium subspecies paratuberculosis (MAP). It commonly affects the small intestine and the colon, often in patches. The infection can cause fistulae to occur between the intestine and other organs. Bowel perforations or colon cancer may also be associated with Crohn's. In the tenth century there were no antibiotics or intensive care unit that would help someone today get through such complications.

There are other diagnoses that doctors have considered for Alfred, for example amoebic dysentery, the disease that likely killed Edward of Woodstock, the Black Prince, a few hundred years later. Alfred's grandson's suffering suggests an inherited, genetic component, however. Whatever the cause, Alfred did a remarkably good job as king, considering the pain and diarrhoea.

The lining of the gut is comprised of epithelial cells. These protective cells turn over quickly and decide what passes over into the blood. A huge array of microbes live here too, the gut microbiome, where a community of bacteria of all different types run the show. The food that we eat has an effect on the microbiome, growing or shrinking the number of bacteria. An area of study known as epigenetics is studying the gene expression patterns in our different gut environments. The gut microbiome affects the body and mind in ways we are in the very early stages of understanding. Over a thousand years after King Alfred the Great struggled with the lining of his bowel, we still only have a fraction of an idea about what really goes on in there. Some might say that the gut microbiome itself should be thought of as an organ, a vital one at that.

In medieval England the bowel was an organ thought worthy of mention, and indeed worthy of punishment during the brutal execution method that was being hung, drawn and quartered, the sentence given to Scottish rebel William Wallace.

Wallace was a leader of the Scots in the late thirteenth century, and is now known to many because of the movie *Braveheart* (even though the name Braveheart was a reference to Robert the Bruce, who we met earlier being lobbed at Moors in Spain). Wallace was a knight during the First War for Scottish Independence. At the Battle of Stirling Bridge, he fought off King Edward I's English army, despite arriving with far fewer combatants. He became a Scottish hero and Edward was not happy. Wallace was captured and in 1305 sentenced to a traitor's death.

As Londoners today go about their business at Smithfield, buying and selling meat, they walk under a plaque, likely without a second glance. Even now, Scots leave heather and symbols of Scotland at this spot to pay tribute to their countryman of 700 years ago.

William Wallace was hanged and drawn and quartered because he had the guts to take on the English, and in reply, the English pulled those guts out in front of him.

Hanging, drawing and quartering didn't happen in that order. First the accused was drawn along the ground. Originally they were tied between two horses, but the victim had a habit of dying too early in the process so later a cart was used instead. They were dragged facedown because they did not deserve to look those watching in the eyes. At the place of execution, they were then lifted high above the ground in 'a place betwixt heaven and earth' because they were not worthy of taking up space in either. They were hanged with a rope around their neck for a while, until just before death. Their genitals were cut off with a knife. This represented that they were undeserving of producing offspring, who would likely follow them down the same path of treason. It was probably a bit late to be reproducing by then anyway.

The shining blade then made a big cut from the sternum and along the abdomen. Through the spurting blood, the guts were pulled out. The bowels represented the wrongdoing that was deep inside them. The executioner pulled out the omentum, the highly vascular cover that sits over the bowel, then the small intestine and large intestine and stomach, whatever they could cut at. There would have been evidence of the dying person's last meal, though it's hard to imagine Edward I feeding his Scottish captive much. The bowels were then set alight in front of the victim and the baying public. Have you ever tried setting fire to slimy human guts? OK, of course you haven't, but it's hard to imagine it is an easy task without a roaring fire. The smell would have wafted across Smithfield to the crowd.

As if pulling out the bowels was not enough, next came the heart and the head, where thought and desire brewed the heinous crime. What was left of the body was chopped up and the parts put on display or sent across the country to warn others of the risks of irritating the King. Wallace's legs and arms were sent north as a warning to the Scots, to Perth and Stirling, Berwick and Newcastle. His bowels didn't make it to those towns, however; with the roads and travel as they were in the 1300s, they would have rotted on the way.

Napoleon Bonaparte's penis

The bloated corpse of Napoleon Bonaparte lay on a cold mortuary table. He was discolouring at the extremities and there was the distinct whiff of decomposition hanging in the air. The last days of the Emperor were painful and sad. He died far away from France, in exile on the remote island of St Helena, as Europe moved on without him. It was May 1821. Napoleon was soon to be buried, but not all of him would make it to his grave.

After years rampaging through Europe, Napoleon was already unwell when he was defeated by Wellington's allied forces at the Battle of Waterloo in 1815. He was suffering from haemorrhoids and pains in his abdomen, but when exiled to the island, his health took a turn for the worse.

An autopsy was performed on the Emperor's corpse with no less than sixteen witnesses. They all agreed that Napoleon was afflicted by a stomach cancer, the same disease that killed his father. Napoleon had had other ideas. Naturally, he believed that he was being poisoned. He had every reason to believe it, with a lot of enemies and a lot of symptoms. His complaints of abdominal pains, headaches, dizziness, nausea, vomiting and limb weakness could all have been the effects of arsenic. As could the night sweats and fevers that plagued him. At times he suffered diarrhoea and at other times constipation. Being at the mercy of his captors, the British, he did not put poisoning past them.

It's possible he was poisoned unintentionally as arsenic was in any number of everyday items used by Georgians in their homes, and in their medicines. If he was sick already, the poison would just make matters worse, exacerbating symptoms and hindering treatment attempts. There were contemporary accusations of foul play, and they continue today. So many people have, over the years, tried to figure out what Napoleon died of, some believing that the British killed their captive. Whatever killed Napoleon, in May 1821 he was just another decaying corpse that needed a burial.

There wasn't really any good reason for what the surgeon did next. Despite being surrounded by so many at the post-mortem, Dr François Antommarchi took up his knife again, grabbed Bonaparte's penis, sliced it off, and hid it from sight. There was a story that the doctor had been bribed to cut off the penis by an unhappy priest, who was upset when Napoleon told him he was impotent. The dish of revenge was served when Napoleon was cold. In this version of events, the penis was handed to the priest, who smuggled it off the island, and Napoleon went to his grave without what is now his most famous body part. What part of the former Emperor would you steal for posterity? Would you take his brain that was once bursting with all those military tactics? His heart, perhaps, that once beat for France? No, neither of those. Let's steal his penis, obviously.

Far more than simply a tube surrounded by columns of spongy tissue, some nerves and an excellent blood supply, the penis obviously had so much symbolic meaning. Napoleon's penis even has its very own Wikipedia page, although at the time of writing this has been earmarked for deletion (pending consultation) by someone with a sensitive disposition.

The priest took his new possession home to Corsica. From there the penis made its way across the sea to England, where it was owned by a Donald Hyde and then a John Fleming, both of them far too interested in the penis, or rather what they could earn from it. London Bookseller Maggs Bros Ltd catalogued it as a 'mummified tendon'. The description worked. The auction gavel came down on the penis lot and it was sold to an American collector called A. S. W. Rosenbach, who was also known as The Terror of the Auction Room and took the artefact to Philadelphia.

Rosenbach was a member of the American Antiquarian's Society and put the body part on display in New York in 1927. On hearing of the exhibition, a reporter at *Time* magazine grabbed his favourite notebook and headed to the Museum of French Art. His excitement to get a glimpse of the one-hundred-year-old penis was short-lived. He was entirely unimpressed by the severed appendage and wrote that it was a 'maltreated strip of buckskin shoelace'. I'm not sure what it was that the reporter was expecting. This was no magnificent display of the Emperor's anatomy, the most famous manly part of a victorious leader who conquered Europe (for a while). Rather, it was the shrivelled unidentifiable relic of a man who fled defeat at Waterloo and died in exile on a faraway island. It wasn't going to win any size contests. But a bad review in *Time* magazine did not end the penis's career. It was then bought for $3,000 by an eminent urologist called John K. Lattimer. His plan was to remove it from public view, stop the spectacle and be more respectful to Napoleon's memory. Now, in the dark basement of a family home in New Jersey, Napoleon's penis lives in a box inside a suitcase.

When Lattimer died in 2007, the penis was passed to his daughter Evan Lattimer, who keeps it under lock and key and will not allow photography or videography. This ensures a little privacy, but it doesn't stop us talking about it. She has said that her father examined the penis. Urologists are specialists in the treatment of the genitourinary tract. If anyone were to know what a real penis looked like, then a urologist would. Lattimer X-rayed it to ensure that it was indeed a penis and not a hoax. It was. There was no way of telling with an X-ray whether this was Napoleon's. This item was small, shrivelled and not particularly penis-looking, but it was a penis none the less.

At first it struck me that this was a very noble thing for the urologist to do. He clearly had an interest in such things beyond ogling them as circus exhibits. Then I learned more about Dr Lattimer and his collection. For a urologist, Napoleon Bonaparte's penis must surely have been the prize possession, but it was not the only grisly historical artefact in his collection.

Amongst the macabre objects on his shelves was the blood-stained clothing of American president Abraham Lincoln. Blood had seeped into the fabric when an assassin's bullet ripped through Lincoln's chest in 1865. Now Lattimer had the collar. He was not planning on washing it.

Dr Lattimer published on the subject of Lincoln's death and the ballistics of his murder. He was asked by John F. Kennedy's family to do the same with the autopsy reports of the latest assassinated president. This time he took a piece of upholstery home for his collection from the seat in Kennedy's car on that fateful day in Dallas in 1963. Goodness knows what that was stained with.

The doctor's collection didn't stop there. He collected medieval armour, cannonballs and rifles, and machine guns from the Second World War. In the 1940s Lattimer worked as a medical officer at the prisoners' barracks at Nuremberg and came home with clothing and underwear belonging to Goering and other Nazis. Hermann Goering had been sentenced to death by hanging but never made it to the gallows. Instead of facing execution, Goering bit into a vial he had squirrelled away containing cyanide. Now Lattimer had the glass vial's brass casing in his collection too. He even took home a piece of rope that was used to hang the Nazi Julius Streicher, and Hitler's X-rays, taken when he was injured in a failed assassination attempt.

So Lattimer was not just interested in Napoleon's penis from an intrigued urologist's perspective, and doing the right thing by it. However, that wasn't the end of the circus around the severed penis, because Napoleon was Napoleon, and because penises are penises.

Napoleon was at first buried on the Island of St Helena where he died. He was later dug up and interred in a grand tomb in Paris at Les Invalides. There have been discussions about the penis being reunited with its original owner in the French capital. There's no proof though that this is Napoleon's penis at all. The French have said that they won't touch it. They 'want nothing to do with the penis', which seems a little out of character for the French.

Though alone in its box in the States, this penis is not the only one to have been removed post-mortem by an interested collector. Grigori Rasputin, lover of the Russian queen (allegedly), was murdered, eventually, by the aristocrat Prince Felix Yusupov and his friends. They tried many ways to rid themselves and the royal family of the Mad Monk. He was poisoned, shot, beaten, and thrown in a freezing river, where he drowned. It was still worth stealing his penis though, even after that. The penis on display in a Russian museum supposedly was detached from his corpse to keep for future generations to get their own glimpse. Whether the rather large organic specimen on display is Rasputin's or not is contentious. Either way, it is a symbol of far more than an ounce of flesh. 'My penis is bigger than your penis,' the Russians shout at the West.

From a small penis in a case, hidden a world away in a New Jersey basement, to a huge, hairy display in yellowing formaldehyde in a Russian museum, we continue our fascination and reverence for the penis. Dr John Hunter was obsessed with two things: being proven right, and the penis. He believed the venereal infections of gonorrhoea and syphilis were the same disease. He decided to prove his theory using his own penis. Hunter took the yellow stinking discharge from a gonorrhoea sufferer and pushed it into little punctures he had made in his own member. He became itchy and had trouble peeing, classic signs of gonorrhoea. Then he got really happy ten days or so later when he developed the classic lesions of syphilis. He thought he had shown everybody that these were the same disease (well, we hope he didn't go around showing everyone); in fact he had managed to infect himself with both diseases.

In Iceland, a whole museum is dedicated to the organ, animal and human. There, rows of specimens, pickled, stuffed and mounted, delight and unnerve us. It claims to be the only place of its kind in the world, which frankly is surprising, considering our ongoing obsession with the phallus. Not everyone is tickled by the displays. For some any representation of the penis is a symbol of toxic masculinity, especially if it is on display in a museum, and we should all just grow up. 'Penis obsessed' and 'power hungry' are two phrases that seem to turn up regularly together.

As for Napoleon's penis, stealing and owning the curiosity was ultimately not about fertility, sex or titillation. It was about power. I wonder, if we searched Dr Lattimer's macabre selection, might we find Adolf Hitler's missing testicle?

Adolf Hitler's testicle

Hitler has only got one ball.
The other, is in the Albert Hall.

The song about Hitler's missing testicle has been heard in the corridors of Britain's schools for over eighty years. The lyrics are sung to the tune of the upbeat 'Colonel Bogey March', as featured in *The Bridge over the River Kwai*, to mock and discredit Hitler (as if he needed any help). Like many a meme, there could be a hint of truth in the song. Is there no smoke without fire, or is this purely a piece of propaganda that has stayed with us for all these years?

Claims about Hitler's genitals, his sexual preferences or acts, regularly pop up and are discredited, only for others to take their place. It's as if Hitler's missing testicle keeps disappearing and growing back. One version of the song goes on to state that Hermann Goering and Heinrich Himmler had micro-orchidism (small testicles) and that Joseph Goebbels had anorchia ('no balls at all'), so perhaps it's not our best source of evidence.

During childhood, testes descend from the body into the scrotum where the sperm can be kept cooler, away from the body's warm core. Undescended testes are the most common birth defect seen in the male reproductive anatomy. The majority will self-correct by six months. Just as we saw with Kaiser Wilhelm II and his withered arm at the start of the First World War, psychologists searching for a motivation for Hitler's atrocities have looked to his body parts. One theory is that Hitler's rampage was in compensation for a condition such as chryptorchidism, which damaged his sense of manliness and self worth. The prevalence of undescended testicles is high and can be an important barrier to psychological development for boys, but not every man with an undescended testis is responsible for mass killings.

Apart from the song, what other evidence is there? In 2008 an eyewitness account popped up supposedly describing the day that Hitler was injured on the Somme. His life was saved by a German army medic named Johan Jambor. Hitler's abdomen and legs were covered in blood, he said, and he had lost a testicle. Jambor claims to have saved Hitler's life. Otherwise he would have been left there to bleed to death. If this were true then it was a significant moment. There he was, an injured Adolf Hitler lying in the mud, covered in his own blood, pulled out to live another day. Historian and biographer Sir Ian Kershaw determined from a military record that Hitler was indeed injured at the Battle of the Somme, but it was a shrapnel injury to his left thigh, not his abdomen. Jambor claimed that, in the 1930s, when the Nazis rose to power, he had nightmares and blamed himself for saving Hitler's life. It's a compelling story.

Two years later, at an auction in Bavaria in 2010, more records came to light. These records were sold but later seized and confiscated by the Bavarian government. When Hitler was arrested after his first attempt to grasp power in the Munich Beer Hall Putsch, he was held at Landsberg Prison. There the medical officer, Dr Josef Steiner Brin, recorded that Hitler had cryptorchidism. This contradicts the idea that he had lost a testicle when wounded by shrapnel in the First World War. In 2015 German historian Peter Fleischmann of Erlangen-Nuremberg University studied the prison papers and claimed it could have been true after all.

There's even a story from Hitler's youth that he had lost the testicle as a child when he tried to urinate into the mouth of a billy goat. Make the accusation sound as silly as possible and maybe more people will spread the rumour.

The Soviets' autopsy supposedly performed on Hitler's remains made no sense. The Soviets said that the missing testicle that they saw (or did not see, as it happens) was on the left. The results were likely propaganda, as the Führer's body was almost completely burnt when found by the Red Army. The tissues were long gone in the fire, and there was no monorchidism to find. They identified Hitler's burnt corpse using a lower jawbone and a dental bridge.

Writer Ron Rosenbaum has tried to have all this talk of Hitler's genitals and sex life put to bed, without much success. After publishing an excerpt from his book *Explaining Hitler: The Search for the Origins of His Evil* in the *New Yorker* in 1995 he received a letter from a psychologist called Gertrud Kurth, who was part of a team attempting to 'evaluate the mind of Adolf Hitler' during the war. They had tracked down a family doctor by the name of Dr Eduard Bloch, a Jewish refugee in New York who claimed that he had examined Hitler during his childhood and found that he was 'genitally normal'. If Hitler had only one ball, then trauma must be the reason why, whether from shrapnel or billy goats, we don't know. One thing we can say, is that the other is not in the Albert Hall.

We often look for physical evidence when someone does terrible things. We want to prove that they're not normal, not like us, not fully human. After his attacks on Ukraine, Vladimir Putin's health has been questioned, and in a twenty-first-century version of Hitler's song, photographs have been widely shared online in which Putin looks sick and on steroid or chemo-therapy treatments. The suggestion is that with such an illness, Putin has nothing to lose. The truth is that the body and its illnesses can only explain so much.

And so back and forth it goes. There will always be questions about the parts of Hitler very few people ever got to see. The-odor Morell, Hitler's personal physician, was known to shout now and again, 'There's nothing wrong with Hitler's testicles.' Does that settle the matter or did he protest too much? The story just does not want to go away. There was even a band in the 1980s called Hitler's Missing Testis, who sang songs addressing the issue. They are hard to find on Spotify but will make your Google search history looking decidedly dodgy.

Henrietta Lacks' cervical cells

When I studied DNA replication, gene expression patterns and immunological cascades during my university degree in the 1990s, I couldn't have known that a lot of what I was learning came from the use of Henrietta Lacks' cervix. HeLa, to me, was just the name of a cell line.

It is not an exaggeration to say that Henrietta Lacks' cervical cells have been the most influential, most shared and most used body parts in history. She never even knew it. Her family knew nothing of it either, at first. Finally, the truth, which was brushed under the carpet for far too long, came to light.

In early 1951 Henrietta Lacks, a Black American woman with five young children, was diagnosed with cervical cancer, a tumour on her cervix. She was born Henrietta Pleasant in

1920 and grew up on a farm that her family had worked for generations, first as slaves but later as free people. In her lifetime, Black Americans would have to use different toilets and water fountains and were forced to occupy different seats on the bus or at the movies, from white Americans.

Henrietta married David ('Day') Lacks when she was twenty years old and they went on to have five children together. Henrietta thought that something might be wrong before the birth of her fifth child. She felt an uncomfortable lump, which the doctors initially thought might be a syphilitic lesion, the result of an infection with treponema pallidum, a spirochete type of bacteria that causes syphilis. When tested, she was negative. So what was the lump? Henrietta was referred to the gynaecology clinic at Johns Hopkins University, the only (relatively) local hospital that would see Black people. There the gynaecologist, Howard Jones, examined her. Henrietta's tumour was malignant, and her worst fears were confirmed. This was a carcinoma, which grew from the epithelial cells, in this case, lining cells that cover the cervix. She was treated with the insertion of radiation rods directly into the tumour on her cervix. During the procedure the surgeon removed tissues for analysis, as is normal, and the biopsies were sent to the lab. Sadly, the treatment was not successful and the cervical cancer killed Henrietta Lacks in October of 1951 at Johns Hopkins Hospital.

Over seventy years after her death, more cells than Henrietta could ever have had in her body whilst she was alive have lived on. Her tissue samples were sent to the lab and labelled *HeLa*, the first two letters from the patient's first and last names. Richard TeLinde, head of gynaecology at Hopkins, was an expert in exactly what Lacks had, cervical cancer. He sent tissue samples to George Gey (pronounced 'Guy') from all his patients, but these tiny cells were like nothing he had seen before. In the lab, Gey noticed that these cells had special properties that meant they could be reproduced in perpetuity.

When Henrietta Lacks died, Gey convinced her doctors that they should do an autopsy. Tissues were taken from other parts of her body too, not just her cancerous cervix. There was no permission asked for, or given, by either Henrietta or her family. The doctors just helped themselves.

Before HeLa cells were discovered and stolen, researchers tried to reproduce all sorts of mammalian and human tissues but they would not live long in the lab. The search was on for cells that would last longer for more in-depth experimentation. Henrietta's cells proved to be everything Gey had hoped for.

The HeLa cell line, as cancerous cells, had mechanisms that prevented them from dying off. Normal cells have a system of checks and balances that prevent them from dividing too fast and replicating too many mistakes. These are the tumour suppressor genes and in Henrietta's cervical cancer cells they had been turned off. There was also the matter of the cells' telomeres, the chromosome end caps. With normal replication, a little of the telomere is lost each time. This redundancy is built into the system; eventually the telomere will be used up and the cell will have to stop replicating. There is a limit then to the number of times a normal cell can divide. HeLa cells see things differently. They make an enzyme called telomerase, which replaces telomere sequences. With this, the cells can divide and keep going for ever. With tumour suppressor genes turned off and with telomerase turned on, Henrietta's cells can live on and on.

Her cells were reproduced in cell cultures. 'Cell cultures' does not refer to what they get up to on a Friday evening, or which art galleries they attend. Cultures are when cells are grown in a lab in an artificial environment. Growing cells need optimal conditions, including temperature and pH, substrates to snack on and a growing medium that provides all they need for growth and reproduction. With the discovery of Henrietta's special cells, researchers could grow them repeatedly and use them indefinitely for their own research projects, or in production of biopharmaceuticals. Because they could be produced without limits, Henrietta's cells could be used to test the effects of drugs, radiation, viruses and toxic substances like never before. They were analysed to understand what cells are made of and what they make. Anyone who wanted the cells to use in their own cellular research could pay a few dollars and have them delivered, anywhere in the world, all without her family's knowledge or financial gain.

The impact that Henrietta's cells have had on the world is beyond calculable. Hundreds of thousands of research papers have been written on the results of the use of HeLa cells. Three Nobel Prizes have been awarded on the back of HeLa cell research. HeLa cells have been the centrepiece of every bit of biological research and breakthrough for the last seventy years. They have helped us understand cells and genetics and been used to develop vaccines and monoclonal antibodies. They have saved lives and made some shareholders very rich. Her family though? Not so much.

Even after Henrietta's death her family were not told the truth. They were kept in the dark for two decades until they themselves were subjected to experimentation without their consent in a bid to create a test for HeLa cells. In 1973 they were called in for blood tests without any knowledge that they were having samples taken for other purposes, and only then did they learn the truth when they were asked for DNA samples. They deserved to be treated better – after all, Henrietta Lacks' cells have given so much to *everyone else*.

As a doctor who studied medical molecular (cell) biology and referenced HeLa cells without any idea of their origin, I can't do Henrietta Lacks and her story justice. It feels too big.

Her body parts, taken without permission, have been used to change the face of medicine for decades, and her family were kept in the dark for years. One of her sons said they were unable to even pay their own medical bills. There were people who made a lot of money selling Henrietta Lacks' body parts. It's a story of racial and class privilege, that shows doctors taking advantage of their patient and once again of body parts not being given due respect after death. Head to toe, lack of respect for other people's bodies is a common theme.

Douglas Bader's legs

Shall we talk about Louis XIV's legs? They were, after all, what killed him. Louis had been complaining of pain in one of his legs for a while. His physicians told him this was a case of sciatica. Sciatica is a pain caused by compression of the sciatic nerve that runs from the lower back down the back of the leg. That's not what Louis was suffering from though. Sciatica, whilst distressing, does not normally turn the skin black and necrotic. Or kill the sufferer.

Something else was causing the King so much pain. No matter what he did he got no relief from the agony. He begged for the offending limb to be amputated and eventually the physicians realised that this could be a surgical problem. But it was too late. Gangrene was killing the tissues and the bacteria had made it to his blood. The Sun King died.

Louis had certainly set a precedent by risking the surgeon's knife before. He went on to celebrate and patronise the work of the surgeons after his successful fistula surgery. He could well have lived a little longer had the surgeons been allowed to remove the problematic limb. It was a very risky operation though, and there's a big difference between removing the roof of a fistula with a scalpel and removing a whole leg with a knife and a saw. Leg amputations were a last resort. It would only be performed if a broken or diseased limb was likely to kill its owner without any intervention. For Louis XIV, or anyone else, to survive an amputation would have been impressive in his century. Amputation, however, was about to become more necessary and a lot more common.

In the sixteenth century, the introduction of firearms brought more complex and severe battlefield injuries. Bullets often dragged foreign matter into victims' bodies. This increased the risk of infection and often resulted in the need for amputation.

On the Napoleonic battlefields of Europe and the civil war battlefields of America in the nineteenth century, many thousands of leg amputations were performed, sacrificing the shattered limbs in an attempt to save the soldier. They had no anaesthetics, but needs must. Henry Paget, Lord Uxbridge lost his leg at the Battle of Waterloo in 1815. He was a cavalry commander, sitting on his horse late in the day when he was hit in the leg by cannon shot. For this officer they bypassed the hospital at the farm of Mont-Saint-Jean and went north to a house in the village where Wellington had his headquarters. Paget was carried to an upstairs room where, without the anaesthetic or antiseptic that we now take for granted, the surgeon cut off what was left of his shattered leg. The broken remains were buried in the back yard and the house's owners even gave the leg its own headstone . . . or should we say, footstone? It became a tourist attraction – *Here lies the amputated leg of a Napoleonic cavalry officer*. There is still a museum on the site now, but it's a bit tired, especially compared to the excellent Waterloo 1815 exhibition built into and under the battlefield which opened for the 200th anniversary.

In 1878, Uxbridge's son visited the site and found that his father's leg bones were no longer buried in the garden, but instead were in a display cabinet in the house. He was furious. He asked for them to be reburied but they weren't.

The story goes that Uxbridge was near the Duke of Wellington when his leg was hit. At that moment he very calmly looked down at the dreadful injury. 'By God, sir, I've lost my leg,' he said, to which Wellington replied, 'By God, sir, so you have.' We can guess what he really said, but that's not stiff-upper-lip enough for a British officer anecdote.

There are plenty of stories about how the officers faired in these battles but little of how the average soldier got on if they had the misfortune to lose a limb at Waterloo. There is very little evidence that wasn't taken for cabbages or sugar. Recently amputated legs were found discarded in a ditch near the hospital, but there was no headstone (or even footstone) for them.

The privileged had access to doctors and the latest in prosthetics. When I visited Castle Fraser in Scotland recently, I came across a leg, in a box in the library. OK, it wasn't a real leg, that would be weird. This was a wooden leg. It caught my eye because it belonged to an officer in the Coldstream Guards in the early 1800s, a man called Charles Fraser who was involved in the siege of Burgos in Spain in 1812. He thought he had got off lightly when he was shot through his hat, but then he was shot again, and again, this time in the leg. He survived but his leg didn't, the musket ball smashed his bones. Injured and sent home, he returned to live at Castle Fraser and had a wooden leg made for every occasion. He was lucky he had the means to do so.

When the castle was taken over by the National Trust for Scotland they found a whole stack of wooden legs in a cupboard that the colonel had had made. They were given to the Woodend Hospital, nowadays ironically used primarily by the orthopaedic surgery department of the Aberdeen hospitals. Though I worked there for a few years, sadly I never came across any more of Charles Fraser's wooden legs. I should have been nosy and opened some dusty old cupboards.

The need for orthopaedic surgeons to perform amputations during wars increased.

Between 2001 and March 2020 there were 302 UK service personnel whose injuries sustained in Afghanistan included a traumatic or surgical amputation. In a similar time frame there were thirty-four UK service personnel whose injuries sustained in Iraq included a traumatic or surgical amputation. The fact that battlefield and hospital care are so effective means that many more are surviving injuries and the loss of limbs that previously would have killed them. They may return as amputees but at least they are returning home. These statistics refer to all amputations, not just legs, as to differentiate between them increases the risk of identification and compromising a patient's right to medical confidentiality.

Amputations have never been simple or easy, and even with all of the medical advancements they remain so today. During the First World War, army doctors were helpless to stop soldiers who lost limbs from suffering in pain. Of the 7 million British soldiers deployed during the Great War, 41,000 had a limb amputated. At the time, chronic pain after amputation was brushed under the carpet. That was partly because doctors just didn't know how to deal with it. They developed a double surgical technique of a quick battlefield amputation followed by a later surgery that tried to prevent the nerve pain brought on by the initial operation.

A century on, improvised explosive devices (IEDs) have made the loss of limbs common among the military once again, but whilst prosthetic technology has improved dramatically, there is still a shortage of effective treatments for pain, especially pain felt in the phantom limbs. World War Two flying ace Douglas Bader must have been through incredible pain when he took to the skies again after a double leg amputation. I watched Bader's 1956 biopic *Reach for the Sky* many times growing up, on account of my father being an RAF officer. It was a black-and-white film starring Kenneth More as Bader. Even as a child I found it wondrous.

Douglas Bader was born in St. John's Wood in London in 1910. His name was pronounced 'Barder', as we learn at the very start of the film when the main character corrects a policeman who stops him for speeding on a motorcycle. It must have been a problem Bader had all his life. He had other, much more serious ones, though you'd never believe it from listening to recordings of him talk, like in his 1981 *Desert Island Discs* recording.

He joined the RAF in 1928 and, following in the footsteps of his older brother-in-law, he went to flying school. He was competitive, athletic and excellent at games, playing outside-half for the RAF rugby squad and at centre for London's Harlequins club. He had been tipped to play rugby for England and he was good at cricket too. As well as games he also loved to fly, loved speed and had a penchant for getting into trouble.

In his own words, Bader had been 'messing about too close to the ground', showing off his acrobatic skills when he clipped the earth with his left wing. Aerobatics were against the guidelines, but then rules were for 'the obedience of fools and the guidance of wise men', Bader said. The aeroplane smashed into the ground and Bader was badly injured. His broken body was pulled from the wreckage and he was rushed to hospital. The surgeons removed one leg that was injured beyond repair and later they had to remove the second when infection set in. As Bader lay in his hospital bed he recalled hearing the voice of a nurse outside his room. She was telling off some other nurses for their noise, disclosing that the man in the room was dying. Those words gave him the strength to carry on and defy her.

Douglas Bader's fitness was the reason he survived. He was up and about within a few weeks and trying prosthetics. He was determined and it wasn't long before he was walking on his new legs. He was ready to get back to what he did best, flying, but the RAF had other ideas. He demonstrated that he could still fly, but there was nothing in the King's regulations about allowing a man with no legs to do so. They turned him down.

Six years later, war broke out and more pilots were needed, so he was able to fly for the RAF again. There were no modifications needed as the RAF aeroplanes were controlled by hands, not feet. Bader became a Squadron Leader and was later promoted to Wing Commander. He had twenty victories in the air but the skies proved very dangerous. During a dogfight Bader's Spitfire was damaged either by shot or when he clipped another aircraft, he couldn't be sure, and he ended up bailing out. He pulled at his prosthetic legs but one was trapped. He had no choice but to leave the right leg behind in the doomed aircraft. He floated to the ground below a parachute with just one prosthetic leg. When he landed he was taken hostage by the Germans. His leg was recovered, and it was battered but usable. After attempting to escape from his hospital bed, he was sent to Colditz. It was, according to Bader, the finest of all the prison camps. He spent two and a half years as a prisoner of war, and unbelievably a new leg was dropped by parachute for him by the RAF at the request of Reichsmarschall Goering. Bader was a high-profile prisoner, but if he did not behave, they could always take his legs away.

After the war, Bader worked for Shell and flew all over the world. The real-life Bader was not quite as jovial as the character More played in the film, but I think we can forgive the flying ace that.

James II of Scotland's femur

Of the 206 bones in the adult human body, the femur, or thigh bone, is the largest and quite hard to break. But not impossible.

A popular online story uses the human femur as a demonstration of the first civilization. Or rather a broken femur. The femur is a big, strong, vascular bone with attachments to equally big muscles. The pain and bleeding of a break would be overwhelming. When an animal in the wild suffers such an injury, the result is catastrophe. The animal is weakened, unable to run, stand and protect themselves. They will surely die. Archaeological evidence of a healed femur shows that someone cared for that individual whilst the injury healed. Therein lies the earliest indication of human civilization.

The story is attributed to anthropologist Margaret Mead, but when another anthropologist, Gideon Lasco, went searching he couldn't find any evidence of Mead telling this story. But that doesn't matter. This story is not about Mead, but rather claims that helping each other through tough times is what makes us civilised. When Mead was actually asked what signalled the beginning of civilisation, she spoke of cities, division of labour and keeping records. That doesn't really pull at the heart strings, does it? Stories need to make us feel something and we can all imagine how breaking a femur would make us feel. To survive, we need someone to care for us. It's nice to know we've been doing that for each other for thousands of years.

Even if someone is caring for us, a broken femur might still lead to death. Bleeding from a femoral break might not be obvious externally but can leak enough inside the body to be problematic.

In the 1740s a ship was wrecked on Anglesey, Wales. Two young brothers were washed ashore and rescued by the locals. One of the boys died and the other was adopted by a local family. The boys spoke no English and were believed to be most likely from Spain, though recent DNA evidence points towards the Caucasus. The surviving boy grew up to be called Evan Thomas and had some remarkable skills. He seemed to instinctively understand how to set bones and had an affinity for using tractions and splinting. Naturally, he became a bonesetter. Bonesetters were not medically trained but their skills were passed down through families. In England there were well-known bonesetter families too – the Taylors in Whitworth and a Matthews family in Warwickshire and the Midlands. Evan's skills were passed on through the Thomas family and his great-grandson Hugh Owen Thomas devised the special Thomas splint in 1865.

Sir Robert Jones, Hugh's nephew, became an orthopaedic surgeon and brought his uncle's invention to the front in the First World War. In 1917, the mortality rates from thigh injuries plummeted thanks to the introduction of the Thomas splint. The next year they fell from 80 per cent to 20 per cent. A hundred years later, a lighter version of the Thomas splint is still used in pre-hospital environments to lengthen and splint a broken femur before the injured can be brought to definitive care.

Thigh muscles are so big and strong that if the femur is fractured the contracted muscles pull the broken bone ends over each other, shortening the length. Bleeding is likely and bones crossed over each other will not heal. Delayed union of the broken bones can lead to permanent disability. When we learn to splint broken bones in first-aid classes we immobilise and strap something hard to the bone so that it doesn't move about. This reduces the pain of broken fragments rubbing against each other. The Thomas splint is a remarkably simple design that goes a step further and saved countless lives. It pushes into the groin above the break and attaches to the ankles below where straps pull the overlapping bone fragments apart to return the bone as best it can to the original shape. As machine-gun fire and shrapnel shot through the soldiers on the front lines of Europe, the Thomas splint was needed more than ever.

At Beaumont-Hamel in the north of France, parts of the trench system still exist in a permanent monument to the battle fought there during the First World War. As I stood on the fire step and looked over the top of the trench I could still make out the shell holes blown into the earth. On that warm, quiet summer's day, it was hard to imagine the horror that took place there in the last century. There were extensive British and Newfoundland casualties here as men were cut down by heavy machine-gun fire. Hot shards of shrapnel too ripped through soft body parts. Among the injured was Private George Henry Prentice. A shell landed outside the parapet and he was blown backwards against the wall. He was knocked unconscious and his thigh was badly smashed. Case notes from his treatment stated that he spent three and a half months 'in splints'. His life and his limb were saved with the use of a Thomas splint, albeit he was left with a slight limp. Rather a limp than a white headstone in one of the CWGC cemeteries dotted across France.

As good as the Thomas splint was, it is not going to save everyone. Some injuries are just too bad. King James II of Scotland's femur was beyond the help of a wire frame and straps.

James was the second in what became a long line of James Stewarts to reign as King of Scots. The first James, King of Scots was chased from his rooms during an assassination attempt at the Blackfriars monastery in Perth in 1437. He woke to hear attackers coming and fled down into a sewer. He was pursued by a gang of conspirators led by his uncle Walter Stewart, the Earl of Atholl. At the end of the sewer James got stuck, there was no way out. A few days before, the King himself had ordered that the sewer be blocked off because he kept losing his tennis balls down it. He had bricked up his own escape route and sealed his fate. The chasers caught up with him and hacked the King to death. His wife, the Queen, was injured but she escaped with her son, who became King James II when he was six years old.

The families around the young king all fought to control him. James saw his fair share of terrible cruelty growing up and it rubbed off on him. He grew up callous and angry. He even stabbed his childhood friend, William Douglas, to death with the help of some friends and threw his body from a window. It is thought James was the inspiration for George R. R. Martin's awful character, King Joffrey in his *Game of Thrones* books.

James II was rather fond of artillery cannons and championed their use. They were the height of murderous technology, destroying buildings and causing general mayhem. He brought Burgundian cannons over to Scotland and used them against the English. In 1460 he was standing next to his huge, prized cannon, during the siege of Roxburgh Castle. It even had a name, Meg. He was showing it off when Meg exploded. Ragged metal sliced through the King's leg and smashed his femur.

The femur's head fits into the pelvis and the bones down to the knee where it joins the tibia and fibula. According to the chronicler Robert Lindsay of Pitscottie, James's thigh bone was 'dug in two with a piece of misframed gun'. The King died quickly, likely from the blood loss. Not only do large blood vessels run down through the legs with the bones, the bones themselves are also highly vascular in that they have a lot of blood vessels running through them. Bones are not just white solid scaffolding that hold up the muscles. The big ones like the femur are also filled with bone marrow, where the blood cells are made, and they bleed. When the metal ripped through the King's thigh, a large proportion of his blood would be running into the field where he lay in a matter of minutes. But it was OK because James had a son. He became James III, King of Scots and in the family tradition he too died a violent death.

Captain Oates' feet

There is a lot going on in the human foot. There needs to be, there's a lot resting on them. Humans are the only large creatures who can continuously stand up on two feet. Lots of bones means lots of joints with lots of ligaments and tendons to keep them all together. With twenty-six bones and thirty-three joints, feet do a valuable job, but they are also vulnerable and can even be expendable.

I will spare you a story about King Louis XIV's gout and his pained, swollen, red-hot and sweaty feet. Jean-Baptiste Lully (1632–1687) was an Italian-born composer of Baroque music who was a favourite of King Louis XIV and his court, and came to the same sticky end as his king.

On this special evening's performance he was conducting a

piece that he had composed to celebrate the King's recovery from surgery, miraculous as it was. Before the modern shorter sticks, the conductor wielded a long staff that would be struck against the floorboards to keep time. He was so enthusiastic for the piece that he stabbed himself in the foot with his conducting stick. Accidentally, of course. He was a violin virtuoso who you'd think might have a little more coordination, but apparently not. He pierced his skin and the spike made contact with the bone beneath.

The tissues of the toe got infected, developed an abscess and turned gangrenous. It was agony as the foot became swollen, pushing at the skin. The blood rushing into the area would make the foot red and hot. Even a blanket would be too much on the painful swelling. The redness made its way up his leg and the pathogens got into his blood until the whole leg was at risk. Lully grew up a dancer and was horrified at the idea of losing a leg, saying he would rather die than have it amputated. At home in a warm bed, Lully's simple foot injury overcame him. He was fifty-four years old.

It was a foot injury that killed the adventurer Lawrence Oates too, but his ending wasn't quite so warm.

Captain Lawrence 'Titus' Oates is remembered as a selfless and quintessential English gentleman who sacrificed himself to save his companions. They did not survive either, but he was not to know that. As a child, Lawrence Oates attended

Eton but he had to leave the college due to ill health. He went on to attend an Army crammer, a school designed to prepare candidates for military life. He was commissioned into the 3rd (Militia) Battalion of the West Yorkshire Regiment. He saw active service during the Second Boer War with a cavalry regiment, but in March 1901 Oates was shot in the thigh. His femur was broken by the bullet and it left his leg an inch shorter than the right. Despite the damage he managed to walk in a straight line and not round in circles. He fought on, refusing twice to surrender and was recommended for the Victoria Cross. Oates was nicknamed Titus, after Titus Oates who had fabricated the Popish Plot in the seventeenth century, but it was a rather ill-fated nickname considering the carnage the original Titus brought to bear.

Captain Oates was made part of Captain Scott's *Terra Nova* expedition to the South Pole because of his experience with horses, of which nineteen were taken along to haul sledges. The £1,000 contribution that Oates made also helped. He was not universally loved or accepted by the team. Belgrave Ninnis, the explorer of the Australasian Antarctic Expedition, called him 'distinguished, simple minded'. The others simply nicknamed him 'the soldier'. Oates clashed with the others, including Scott. He wrote that Scott's 'choices of animals were off' and 'he was not a great organiser of animals or men'. According to Scott, 'the Soldier takes a gloomy view of everything.'

In January 1912 the five-man team reached the South Pole. They found a tent left there by Roald Amundsen, the Norwegian explorer who had beaten the British team to glory by over a month. With that disappointment hanging over them, the Brits now faced the long journey home. Conditions were terrible with bad weather, injuries, scurvy and frostbite all a threat. One of the team, Edgar Evans, died from a head injury suffered in a fall a few days earlier.

As the team walked back towards their base, the extreme cold got to Oates. He developed terrible frostbite that froze his feet and gangrene set in.

When exposed to extreme cold conditions, the extent of the damage is proportional to the falling temperatures and the duration of exposure. They were out in the cold for a long time. When the body is exposed to below −2.2°C, ice crystals start to form. I can feel the tingling and pain just typing that. This occurs both within the cells and in the extracellular spaces. The ice crystals cause the cells to break open, shattering the outer membranes. Frostbite injuries can occur in two major ways. The first is the initial freezing of body tissues. The other is the reperfusion injury that occurs when the frozen tissue thaws. When this happens within blood vessels the ice crystals cause damage to the endothelium, causing stasis, clotting and tissues to die. Fluid shifts occur and so oedema makes the area swell. The little blood vessels collapse and tissue necrosis turns the area black as it dies.

If you have ever tried to freeze strawberries, you will see that they never thaw to quite their original form. When water freezes, it expands. When water freezes within the cells of the strawberry, the cells burst open and the strawberries will be a mushy mess. Cells of the body, the fingers and toes, and nose, will also burst open and go mushy if they are frozen solid.

At first Oates would have felt tingling in his toes, with itching and irritation. Then blisters appear on the skin. The layers of the skin freeze one by one and worst of all, the tendons and bones within his feet froze. A long way from the heart and the organs that require constant warm blood supply, the feet are vulnerable. The body sacrifices circulation to the feet when a choice must be made between life and limb. The body has a remarkable mechanism whereby the infected or injured body can cut off appendages if they are dead from frostbite or gangrene. These limbs are not vital to survival but their gangrene is potentially harmful. This is known as auto-amputation, the breaking off of blackened necrotic tissue, and whilst it could cause disability, it can save lives. For Oates, it was far more serious. This was not just one or two exposed toes, but his whole feet.

On 15 March, Oates told his teammates to go on without him. His walking had slowed; he knew he was holding them back. But they refused to leave him behind in his sleeping bag as he asked.

On 17 March, Oates knew he had to do something to allow his team to go on. He felt the only thing he could do was hobble off into the storm. He got up onto his frozen feet and walked out of the tent into a blizzard at −40°C to die on Antarctica's Ross Ice Shelf. 'An act of a brave man and an English gentleman,' Scott wrote in his diary, where he also recorded Oates' last words to them. 'I am just going outside,' he said, 'and may be some time.' It was his thirty-second birthday and it was his last. Perhaps this book needs a chapter dedicated to the stiff upper lip.

Captain Lawrence Oates was not a typical young man of his day. He fought against societal norms because he felt so uncomfortable with their expectations. Neither contemporary customs nor clothes seemed to fit with Oates. It's often men who do not fit in that are attracted by the call of adventures, of extreme exploration, of being thousands of miles away from everybody else on the planet, and aiming for lofty, perhaps impossible goals. He was also a long way from his mother, who was a rather overpowering character. It was she who first questioned the whole ill-fated expedition, looking for answers when her son did not return home.

There has been plenty of examination of Oates' story over the years, and with it plenty of mudslinging, including a story involving Oates getting a young girl pregnant before running off to the South Pole. It was Oates' diaries, bemoaning Scott's lack of leadership abilities, which seemed to contradict the official story of gallantry, causing some to lay blame on Scott or even Oates himself for their failure. This story, however, is about his feet, and they froze on the Ross Ice Shelf in 1912. There is a commemorative plaque at the Holy Trinity Church at Meanwood in Leeds, near the Oates family home, dedicated to Captain Oates. It rather poetically states: 'His body lies lost in the Antarctic snows.'

During the First World War, soldiers were shown pictures of Oates and his South Pole companions as examples of how noble men of Britain faced death. In the trenches of Europe, under heavy fire, soldiers were asked to draw strength and courage from the story of Oates keeping a cool head in the face of tragic circumstances.

Frostbite hit the headlines recently when Prince Harry, the King's youngest son, wrote about his experiences adventuring in the South Pole. He felt the onset of frostnip on his todger (the rather posh British term he used for his penis to save blushes). Luckily for Harry, his problem was just a small one. His penis wasn't at risk as long as he warmed up.

On the other end of the spectrum of cold-weather injuries is trench foot. The soldiers' feet in the trenches faced a lot worse than Prince Harry's todger. Trench foot is thought to be caused by successive vasodilation and vasoconstriction, the opening and closing of the blood vessels in response to freeze and thaw.

Skin and soft tissues break down, increasing the chance of infection getting in. The legs can swell to the knee and blisters full of foul-smelling fluid form. It's an odour you won't soon forget. Trench foot tends to be associated with the trenches of the First World War, where it was the scourge of many a soldier, but can happen anywhere the weather is as bad. Unlike frostbite, trench foot does not need the freezing temperatures to cause damage and so can be seen in temperate environments. In waterlogged, cold trenches, foot hygiene was crucial but not simple. Dominique-Jean Larrey, the famed French doctor favoured by Napoleon, wrote of how such a disease affected camps of soldiers a hundred years previously, though it did not take the name 'trench foot' until the Great War, where in 1914–15 alone it was estimated that 20,000 British troops were treated for the condition.

Yao Niang's toes

In the name of fashion and in the search for a better life, millions of young girls took part in the Chinese practice of foot-binding, which lasted for a thousand years. The long-told and likely embellished origin story has it that in the tenth century, just before the time of the Song dynasty, emperor of China Li Yu had a favourite concubine called Yao Niang.

One day she wrapped her delicate small feet with silk and danced beautifully to depict a golden lotus. It was a stunning display that took the breath away. So many wanted to emulate Yao's beauty that the women started to bind their own feet so they could squeeze into as small a size as possible. The smaller the better.

There are no other details about the first dancer that birthed the fashion, so it is no surprise that the story would be changed, added to, altered and re-imagined over so many centuries.

At the age of about four years old, girls went through their first ceremony of foot-binding. The feet were soaked in water and rubbed with herbs and oils, and the toenails cut. Then the binding would start. The big toe was left untouched but the other four toes were subjected to a painful deforming process. The joints holding the bones of the toes together were broken and wrapped underneath the foot, ultimately bringing the foot to a triangular shape. The cotton wraps were regularly removed and the feet cleaned of blood and pus before being tightened further. Toes were beaten, rubbed and manipulated into the desired shape. The length of the foot was reduced further as walking broke the toes again, constricting growth and pushing the foot into a pronounced arch.

The transverse tarsal joint, which brings the navicular and cuboid bones to the calcaneus (the heel bone), is broken apart during foot-binding. In the forefoot, each toe is made of four bones, first the metatarsals and then three phalanges decreasing in size as you move away from the ankle. The small tendons and ligaments which hold the bones together and allow walking are destroyed in this process. The normal angle from underneath the foot to the instep is between 20 and 30 degrees. In a foot-bound woman, the angle reaches 60 to 80 degrees. The heel bone is shifted to near vertical. The process creates a large cleft under the foot. The inability to walk reduces the bone density in the hips and spine, meaning any fall risks fracture, and tiny feet might make falling over more likely.

The aim was to mould the foot into the Golden Lotus, a foot that was only three inches long. A foot that achieved four inches was the Silver Lotus. A foot that had gone through all the pain but only reached a length of five inches was known as an Iron Lotus.

But why? Emulating a delicate dancer is one thing, but why did it stick for so long? Small feet were considered the height of elegance and sophistication. At first it was a status symbol for the higher classes during the Song dynasty (960–1279 CE) and, as is usually the case, others followed the upper classes. The lotus shape of the foot was also hugely fetishised, as was the style of walking which had to be adopted with bound feet. It was seen as attractive, erotic even. Small feet became associated with moral virtue and modesty. What's more, the smaller the feet, the more likely a young girl was to make a good marriage. It was the duty of the mother to help, to insist on the bindings as an act of love. They believed it would give them a better future, and it probably did, as men looked for tiny-footed women.

However, foot-binding was also used for the control and subjugation of women. Once the feet were bound, the girls could do far less physical work and were forced to do sitting occupations such as needlework. They could not leave the house or travel far and were therefore easily controlled. It was an intensely painful process that risked infection or even the loss of toes (though that was considered an advantage as there was less foot to deal with). Eventually the women were largely left without pain as their feet became numb.

In the seventeenth century the Emperor banned binding but the edict was mostly ignored. Later Jesuit missionaries tried to have it banned without much luck. In 1888 Kang Youwei founded the Anti-Footbinding Society to combat the practice. In 1912, this thousand-year practice was finally outlawed, although it still continued in secret. The last recorded incidence was in 1957, four decades after its banning.

The Zhiqiang Shoe Factory in Harbin that specialised in the tiny shoes worn by the foot-bound women of China, closed its doors in 1999. Their now ageing clients were dying and there was no longer a need for the lotus shoes. Only a handful of women who went through the process remain. They have daughters, granddaughters and even great-granddaughters who will not have to go through the suffering as so many generations before them. Foot-binding is now seen as a symbol of old China, and many in the futuristic cities of high-rise buildings would rather see it forgotten altogether.

We know much of this because the photographer Jo Farrell went searching for the last-living women, then between eighty and a hundred years old, whose feet were bound. They allowed Jo to photograph their feet and told her their stories.

Whether there really was a Yao Niang who started the fashion or not, it is remarkable that the process spread and survived all over China for centuries. The attempt to emulate the beauty of one dancer caused millions of Chinese girls to alter their feet a very long way away from what their DNA intended.

Freddie Mercury's white blood cells

It's a shame we can't see our white blood cells doing their job. They really do look after us. That is, as long as they do not encounter an adversary like the human immunodeficiency virus (HIV). If the white cells are broken by an infection with the virus, there is potential for disaster.

Freddie Mercury was born Farrokh Bulsara in September 1946. He was the lead vocalist of Queen, the frontman of a very special band and a massive worldwide hit. In 1985, Mercury and the band took to the stage at Band Aid raising funds for charity. The telephone donations, which had been slow at first, suddenly came flooding in. It was an iconic moment that felt like the whole world was watching. What they could not see on their TV screens was that inside Freddie's white blood cells, trouble was brewing.

The singer's wild promiscuous rock and roll lifestyle had caught up with him. He was infected with the human immunodeficiency virus. Transmitted from person to person via sex, blood products or the sharing of needles, the virus was spreading, primarily among the gay community of which Freddie was a part. Fear and scaremongering made everything worse. Even testing for HIV came with a stigma that prevented those who needed it most from accessing care and resources. The blood test for HIV was available from 1984 but it was not just stigma that stopped people taking the test. Many felt it was pointless to take a test for a disease that couldn't be treated, like HIV. Once HIV became AIDS, it was a death sentence.

The tiny virus structure gets itself into the blood of a new host and targets the immune cells. HIV belongs to a class of viruses known as retroviruses. These viruses use RNA (ribonucleic acid) to encode their genetic information rather than DNA (deoxyribonucleic acid). They work in the opposite direction to a human cell where the DNA message is sent through RNA listing the building blocks that make the protein. Hence the name, retrovirus. In RNA, the base thymine is replaced with uracil. There's not a lot to this deadly virus. The twenty-faced polyhedron contains a capsid with two strands of RNA, with a single helical band of nucleic acids inside. It's a simple yet destructive little thing. On the RNA strands there are nine genes containing all the instructions needed to make new viruses. Spikes of glycoproteins fit within a lipid outer layer and when they grab on to the CD4 receptors on the body's T-cells, the virus gains entry into the cell. White blood cells, leukocytes, are made in the bone marrow and are found circulating in the blood and the lymph tissue. They are part of a cell-mediated immunity system that the HIV virus takes advantage of. If the T-cell has a CD4 spike the virus can find them.

Untreated, the virus goes from acute infection, a seroconversion, where some feel like they have the flu, through to chronic infection which is at first asymptomatic. The virus at this later stage is infecting new cells, replicating and starting to cause damage. The chronic stage then becomes symptomatic. That's when Freddie started to feel unwell and symptoms appeared. With Freddie's white cells being attacked by the virus, serious infections or cancers that under normal conditions the immune system can repel were free to grow. He found a Kaposi's sarcoma on his shoulder, a reddish-purple lesion that was a telltale sign of an immune system weakened by HIV. Kaposi's sarcoma is caused by a member of the herpes virus family that only causes trouble when the immune system can't stop it. These highly visible lesions are sadly a stigmatising sign of the disease. The HIV virus stops the immune cells from doing a job we almost take for granted. With the T-cells knocked off, infections can roam free. The late stage of HIV infection, acquired immune deficiency syndrome (AIDS), can be identified by diseases that are otherwise seen very rarely.

When Freddie eventually decided to find out if he had HIV, news of his test leaked to the press. By the mid 1980s there were thousands of gay men dying of AIDS. Freddie's test came back positive too. He was carrying the virus and his white blood cells were struggling to function. He didn't tell anyone but kept on working. The great pretender.

At the BRIT Awards in 1990, Freddie looked gaunt, a shell of his former vibrant self. In 1991, he released his last video singing us his last song. He looked like there was nothing left of him and yet he sang 'These Are the Days of Our Lives' directly down the camera at us, with his trademark Freddie passion. Filmed in black and white to hide the full extent of his illness, it was clear he was very sick. He died of AIDS, or more specifically the disease bronchopneumonia that AIDS prevented his body from fighting off, six months later, on 24 November 1991.

As if the devastating disease needed any more awareness, Freddie's death certainly kept it fresh in people's minds. The effort to improve the lives and outcomes for those with the virus was immense. Molecular biology techniques were being improved upon in labs throughout the world in the latter part of the twentieth century, and with that came a better under-standing of the retrovirus. The virus has not gone away since then with 38.4 million people globally living with HIV in 2021, but now more treatments can be offered because of all that effort. Were Freddie diagnosed with HIV today, his white blood cells would stand a much better chance. The development of antiretroviral medications means that people living with HIV can keep the viral load below unwelcome levels and their own CD4 T-cells working to fight off other infections. As Freddie told us not long before his death in 1991, the show must go on.

Remains to be seen

When medical schools teach human anatomy and physiology, they have a habit of separating the body into parts or systems. We can easily forget that all the body parts come joined together, attached by connective tissue, the stringy sinews and skin. In Portugal, the phrase 'Inês is dead' is used to mean 'It's over, give up, stop digging things up'. We can't seem to take the advice though. Humans love to dig up the dead.

Before Pedro I of Portugal became king in 1357, he fell madly in love with one of his wife's ladies-in-waiting, her cousin, the Galician noblewoman Inês de Castro. They became lovers and had illegitimate children together. Pedro was smitten and so wasn't that bothered when his wife died and left them to it. His father, King Afonso IV, was furious that Pedro and Inês' relationship continued. He had strictly forbidden it and had been ignored by his son. They argued over titles, money and everything else. Afonso eventually sent a murder squad to find Inês and they hacked her head off. Pedro was understandably distraught.

The first thing he did when his father died and he became king was to take violent revenge on his lover's murderers. Then he dug up Inês' body and declared her two-year-old corpse queen. He had forgotten the 'til death us do part' bit in his vows. They were married, he said, and that legitimised their children. He dressed her body in royal robes, crowned her rotten head and put her on the throne. A parade of onlookers were made to kneel down before the dead queen and kiss her rotting hands in allegiance. But he wasn't the only one to kiss the corpse of a queen.

On his thirty-sixth birthday our old friend Samuel Pepys, still free of pain after his bladder stone operation, celebrated his birthday with a visit to Westminster Abbey. It was Shrove Tuesday, he should have been at home eating pancakes, but no, Pepys went to the Abbey to see the body of the long-dead Queen Catherine de Valois. Catherine was the wife of King Henry V and so lived two centuries before Samuel Pepys. For a fee, visitors to the Abbey in London could see her corpse, which had been removed from its resting place and not yet put back. Pepys wrote in his diary, 'Today I did kiss a queen.'

These two queens weren't the only ones dug up and paraded. Before crossing the Channel to become the conquering King William I of England, William of Normandy tried to rally his morale-sunk troops by digging up the body of St Valery and parading him. St Valery was a French monk who died in 619 CE, credited with numerous miracles that illustrated his life of prayer and sacrifice. It must have helped. William became the Conqueror.

Years earlier, the Cadaver Synod was held by Pope Stephen VI, who had been bitter enemies with Pope Formosus. Nine months after the death of Formosus, Stephen exhumed the body. He put Formosus on trial, charged with perjury. When he didn't answer any of the questions put to him, a deacon was made to speak for the corpse. Presumably he sat behind the dead pope projecting his voice and controlling the cadaver like a puppet. Unable to defend himself, Pope Formosus was found guilty. His fingers of consecration were cut off, he wouldn't be needing them any more, and his papal election was declared invalid. He didn't even respond to the verdict. Cold.

Enough people found the whole saga bizarre enough to rise up against Pope Stephen and strangle him to death. The pope that came next, Pope Theodore II, put everything right again, vindicating Formosus. Theodore himself only lasted twenty days as pope. Some think those who bore a grudge against

Formosus took it out on Theodore. There was no resting in peace.

Putting dead bodies on display has not been reserved for queens, saints and popes. There are now hugely successful exhibitions that display whole dead human bodies. Not in the name of religion or power, but in the name of education.

Last year, I was wandering around London and came across posters for Gunther von Hagens' Body Worlds exhibition at Piccadilly. Staring out at me was a human head, stripped of skin, with muscles and tendons stretched across a skull. I had almost forgotten about this exhibition after the huge hype it had in the late 1990s, a time when I was dissecting bodies myself in anatomy classes.

Von Hagens is an anatomist who achieved a doctorate in Heidelberg in the 1970s. He's a showman who is intentionally shocking and always wears his trademark fedora. It's a literal hat-tip to Nicolaes Tulp, the subject of Rembrandt's painting *The Anatomy Lesson of Dr Nicolaes Tulp*, who wears such a hat. In the 1990s von Hagens appeared on a UK talk show presented by Jonathan Ross, emerging as a band sang 'The Monster Mash'. Ross introduced him as the man who makes works of art from human corpses.

Von Hagens was a doctor who developed his techniques with a mind to education. He even described himself as a 'public anatomist'. He designed a new plastination process that, he said, brings the tissues to a place between death and decay. Back in the 1970s, he saw a specimen that had been encased in plastic in order to preserve it. He had the idea to put the plastic on the inside of the specimen instead. He then spent two decades perfecting the technique he called plastination. The water was removed from all of the cells, to be replaced by acetone that boils at a low temperature, so that is then vapourised, and with a vacuum process the polymer is drawn into the cells. Polymers harden and remain in the shape of the tissue they had filled.

Von Hagens believed aesthetics should be considered so that doctors could take pride in their specimens, making the anatomy displays emotional rather than just placed on their backs, hands by their sides, palms facing the ceiling. After all, Renaissance anatomists put bodies in lifelike poses and displays, influencing von Hagens as we see with his trademark hat. The anatomists of old drew and sculpted depictions of the body in the écorché style, showing the muscles and organs without the skin. Van Hagens' exhibitions take inspiration from the traditional seventeenth- and eighteenth-century style, the only difference was he used real people and not sculptures.

Von Hagens attracted controversy by performing an autopsy live on Channel 4. It was the first public dissection for well over a hundred years. There was as much appetite for it this time round as there was in the nineteenth century. After that von Hagens took his dead bodies on tour. Body Worlds was the exhibition arm of Gunther von Hagens' body of work (sorry about that one). The show, originally housed at Brick Lane, contained various anatomical specimens, from small blocks of preserved tissue to full posed cadavers in varying states of dissection and preservation. Full-body plastinates can be seen in various poses. Smaller specimens show the organs of the body in various states of disease or health. Visitors are greeted by a skeleton at prayer, holding up a human heart. There are shiny prosthetic implants sitting within joints – a sight usually reserved for the orthopaedic operating theatre or the autopsy table.

As the spectacle grew in the late 1990s and early 2000s, so did von Hagens. He opened a lab in Dalian city in China with the help of a Chinese doctor, a graduate of Dalian Medical University called Sui Hongjin. The Body Worlds exhibition became an enormous enterprise. At one point there were 240 employees working at five different labs in three different countries, Germany, China and Kyrgyzstan. If something is successful and making a lot of money, it won't be long before it is copied. Sui Hongjin later used the plastination process he had learned from von Hagens and set up a rival attraction, BODIES: the exhibition.

Sui Hongjin was based in China, where people generally don't donate their bodies to medicine or science, and yet Sui was able to get his hands on them. It was thought that the bodies were coming from the police, but where the police acquired them is unknown. Alarms bells were ringing and the German magazine *Der Spiegel* published an exposé. They claimed that at least two of the bodies in von Hagens' stories had bullet holes in the head. It was a serious accusation. Perhaps, the newspaper claimed, they were even prisoners from the banned Falun Gong religious movement. Von Hagens was furious. He denied using Chinese cadavers in his exhibitions, and insisted that his cadavers were all ethically sourced, and in 2003 he rightly returned seven corpses to China that were of questionable origin. He got an interim injunction preventing the magazine publishing such allegations about him again.

It wasn't *just* Chinese bodies that had caused a commotion. Earlier, in 2002, two Russian doctors were charged with the illegal supply of fifty-six bodies to von Hagens. They were accused of supplying the dead bodies of convicts, the homeless and the mentally ill without any consent from them or their relatives. Once again we find ourselves embroiled in an argument about consent. Von Hagens was clear that the bodies were sent to his lab for him to carry out plastination and return them to Russia, being the contractor, if you will. He confirmed that the bodies from those doctors never appeared in any of his exhibitions.

Again he said that those bodies on display in his Body Worlds show had full consent. He also had a long waiting list of people who were keen to donate their bodies. With a long list of people willing to donate their bodies, why would he need to pay for bodies of questionable provenance?

When a different human plastinates show came to his city in 2007, the Bishop of Manchester called the exhibitors Body Snatchers. He was referring to the original Body Worlds but also its copycats: Bodies: the exhibition; Our Body: The Universe Within; Bodies Revealed and Mysteries of the Human Body. These later shows took the plastination techniques without Dr von Hagens' original educational ideas and moral reasoning. Like him or not, the doctor could explain what he was doing, and why he was doing it, unlike many of the later shows that were solely freak-show voyeurism. And what's more, some of the copycats' bodies *did* come from China.

Lord Alton of Liverpool spoke in the House of Lords in 2021, saying that there 'were probably people who had been executed in China' featured in one of the copycat shows at the NEC in Birmingham in 2018. New York Attorney General Andrew Cuomo, after investigations that also looked at organ-harvesting, said, 'despite repeated denials, Premier [who are the company behind the other exhibitions] can't demonstrate the circumstances that led to the death of the individuals'.

I don't know about von Hagens either way. I like to think

that underneath the slightly sinister showman surface is a man who simply loves human anatomy and wants to share it with us through plastination, even if his showmanship has probably also served his bank balance very well. The same can't necessarily be said for the non-Body Worlds affiliated shows. They remain popular and continue to highlight the consent debate, like the one surrounding the centuries-old bones of the Irish giant, Charles Byrne.

Some believe displaying bodies to a paying public is wrong, regardless of consent. Religious ethicist Thomas Hibbs said that by manipulating body parts in this way the exhibitors were stripping away human significance. This was dead-body porn, he said, questioning why many millions of people queue to see bodies. The practice may go against Hibbs' own personal beliefs (religious or otherwise), but it's not necessarily against society's best interests if conducted in the right way. These shows provide a view of the human body that is simply not possible for those who do not work in healthcare. With the current sterile way we medicalise death, many have never even seen a dead body, so these shows have a chance to both educate and help familiarise us with death before we face our own or that of a loved one.

Back in London, when I stumbled across the poster, any excitement about seeing the exhibits again was short-lived. The posters were faded and frayed at the edges and the venue was in a bad state. The doors were shut, the alleyway to the entrance reeked of urine. I had not realised how big Body Worlds had become. Millions had seen it at its new home in the centre of London where the Ripley's: Believe it or Not! exhibition had been, but it did not reopen after the Covid lockdowns of 2020. I don't know if the London Body Worlds exhibition intends to open the doors again. They have a lot of work to do if that is the plan – someone needs to clean that entrance alley for a start. Von Hagens is in his seventies and sadly has been diagnosed with Parkinson's disease. He surprised his wife with a request that his body be put through the plastination process and put on display to greet visitors to the exhibition, complete with his trademark hat. I would go and see that. What does that say about me? I will leave you to decide.

Millions of people, all over the world, have queued, paid for and filed past the global phenomenon that is the Body Worlds exhibition. For science? For learning? For voyeurism? Well, it's not for the squeamish. For me, it should be for each individual to decide. Censorship is for religions and dictators. I for one have no problem with the display of human anatomy with the correct consent for the purpose of education, bearing in mind that such exhibitions are educational even if they are not affiliated to university courses or science centres, because they expand the experience of life.

As for my body? Do with it what you like. Bind a book,

vital organs

display my bones, dissect me to learn anatomy, stick my head in a box or stick my middle finger up at everyone. I'm not fussed. It could be educationally helpful to someone else, but I don't believe it will affect me, wherever I'm going next.

The (rear) end.

Selected Bibliography

Ashcroft, F., *Life at the Extremes* (Flamingo, 2000)

Bierman, Stanley M., *Napoleon's Penis: Plus Other Engaging and Outrageous Tales* (Trafford, 2012)

Brickhill, P., *Reach for the Sky* (Collins, 1954)

Crumplin, M., *Guthrie's War: A Surgeon of the Peninsular & Waterloo* (Pen & Sword Books, 2010)

Duffin, J., *History of Medicine* (MacMillan Press, 2000)

Fong, K., *Extremes: Life, Death and the Limits of the Human Body* (Hodder, 2013)

Fitzharris, L., *The Butchering Art: Joseph Lister's Quest to Transform the Grisly World of Victorian Medicine* (Penguin Random House, 2017)

Gaudi, R., *The War of Jenkins' Ear: The Forgotten Struggle for North and South America, 1739–1742* (Pegasus, 2021)

Ghossain, A. and Ghossain, M. A., 'History of mastectomy before and after Halsted', *Le Journal Médical Libanais*, Apr–Jun 2009, 57(2):65–71

Gillman, P. and L., *The Wildest Dream: Mallory: His Life and Conflicting Passions* (Headline, 2000)

Gordon, R., *Great Medical Disasters* (Hutchinson & Co., 1983)

Herrera, H., *Frida: A Biography of Frida Kahlo* (Harper Collins, 2022)

Hume, M. A. S., *Spain: Its Greatness and Decay, 1479–1788* (Cambridge University Press, 1905)

Kean, S., *The Tale of the Duelling Neurosurgeons: The History of the Human Brain as Revealed by True Stories of Trauma, Madness and Recovery* (Penguin Random House, 2014)

Keith, A., 'The Skull of Lord Darnley', *British Medical Journal*, September 1928, p. 458

Ko, D., *Cinderella's Sisters: A Revisionist History of Footbinding* (University of California Press, 2007)

Lepore, F. E., *Finding Einstein's Brain* (Rutgers University Press, 2018)

Loveman, K. and Pepys, S., *The Diary of Samuel Pepys* (Everyman, 2018)

Lustig, R., *Fat Chance: The Hidden Truth about Sugar, Obesity and Disease* (Fourth Estate, 2013)

Michals, T., *Lame Captains & Left-Handed Admirals: Amputee Officers in Nelson's Navy* (University of Virginia Press, 2021)

Mickle, S. F., *Borrowing Life: How Scientists, Surgeons, and a War Hero Made the First Successful Organ Transplant a Reality* (Charlesbridge, 2020)

Moore, W., *The Knife Man: Blood, Body-Snatching and the Birth of Modern Surgery* (Bantam Press, 2005)

Obenchain, T. G., *Genius Belabored: Childbed Fever and the Tragic Life of Ignaz Semmelweiss* (University of Alabama Press, 2016)

Richards, M. and Langthorne, M., *Somebody to Love: The Life, Death and Legacy of Freddie Mercury* (Weldon Owen, 2018)

Skloot, R., *The Immortal Life of Henrietta Lacks* (Macmillan, 2012)

Stark, P., *Last Breath: Cautionary Tales from the Limits of Human Endurance* (Pan Books, 2001)

Summers, J., *Fearless on Everest: The Quest for Sandy Irvine* (Mountaineers, 2000)

Tilney, N. L., *Transplant: From Myth to Reality* (Yale University Press, 2003)

Tolet, F., *A Treatise of Lithotomy: or, Of the Extraction of the Stone out of the Bladder. Written in French by Mr Tolet. Translated into English by A. Lovel* (1689)

Tomalin, C., *Samuel Pepys: The Unequalled Self* (Penguin, 2003)

Yudkin, J., *Pure, White and Deadly* (Penguin, 1986)

'A revolution in treatment: the Thomas splint', The National Archives blog (accessed 03/07/2023)

'Emily Davison and the 1913 Epsom Derby' The National Archives blog (accessed 05/01/2023)

Letter from Frances Burney to her sister Esther about her mastectomy without anaesthetic, 1812, British Library (bl. uk) (accessed 03/07/2023)

'Suffragettes: Accident involving His Majesty's horse and jockey, The National Archives, Ref: MEPO 2/1551 Description: Suffragettes: Accident involving His Majesty's horse and jockey (accessed 03/07/2023)

Acknowledgements

Vital Organs has been so much fun to research and write. Once again I owe much to Emily Glenister, my agent at DHH Literary agency, who provides endless enthusiasm for my human body stories and continues to guide me through the world of publishing. Thank you also to David Headley and the staff of DHH and Goldsboro Books who are always so welcoming when I travel to London.

At Wildfire, Philip Connor has provided reassuring guidance throughout and I appreciate that so much. Thank you to Areen Ali, Emily Patience, Isabelle Wilson, Jill Cole and the extended team at Wildfire and Headline. Hayley Warnham has created a fabulous blood-spattered cover design.

To my medical friends The Elites, and the Orleton Park gang

who remain fearless cheerleaders, thank you for being there, I hope you enjoy my stories of body parts from history. Sadly, none of your amusing title suggestions made it through but thanks for trying.

Thank you to my wonderfully enthusiastic followers on TikTok. The numbers grow and grow. Ryan Clark, Melissa Ratliff and Jennifer Wilcox cheer me on from afar and Kristin's postcards have kept me entertained and motivated. The two Marks and everyone in the BXP team have been hugely supportive, understanding and encouraging.

To the Edges: Derek, Kathryn, Charlie and June, and to the Thompsons: Liz, Guy and Andy, thank you again.